RENEWABLE ENERGY: RESEARCH, DEVELOPMENT AND POLICIES

MASTER LIMITED PARTNERSHIPS AND REAL ESTATE INVESTMENT TRUSTS

RENEWABLE ENERGY PROPOSALS

RENEWABLE ENERGY: RESEARCH, DEVELOPMENT AND POLICIES

Additional books in this series can be found on Nova's website under the Series tab.

Additional e-books in this series can be found on Nova's website under the e-book tab.

RENEWABLE ENERGY: RESEARCH, DEVELOPMENT AND POLICIES

MASTER LIMITED PARTNERSHIPS AND REAL ESTATE INVESTMENT TRUSTS

RENEWABLE ENERGY PROPOSALS

JOSIAH HESLIN
EDITOR

New York

For permission to use material from this book please contact us:
Telephone 631-231-7269; Fax 631-231-8175
Web Site: http://www.novapublishers.com

NOTICE TO THE READER

Library of Congress Cataloging-in-Publication Data

ISBN: 978-1-63117-012-6

Published by Nova Science Publishers, Inc. † New York

CONTENTS

PREFACE

Topics discussed in this compilation include the opportunities and potential complications for renewable energy with the master limited partnerships and real estate investment trusts; the technical qualifications for treating photovoltaic assets as real property by real estate investment trusts; master limited partnerships parity act; and master limited partnerships as an option for the renewable energy industry.

Chapter 1 – Master Limited Partnerships (MLPs) and Real Estate Investment Trusts (REITs) are two proposed investment vehicles that have the potential to lower the high cost of capital for renewable energy assets—a critical factor in the U.S. Department of Energy's goal for renewable energy to achieve grid-parity with traditional sources of electric generation. Due to current U.S. federal income tax laws, regulations, and administrative interpretations, REITs and MLPs cannot finance a significant portion of the cost of renewable energy assets. Efforts are underway to alter these rules by changing the definition of "real property" (REIT) and "qualified income" (MLP). However, even with rule changes, both investment vehicles have structural challenges to efficiently finance renewable energy assets. Among them are (1) effectively utilizing the U.S. federal income tax incentives; (2) administratively structuring the investments to not be overly onerous or complicated, given the potential for pooling a relatively large amount of small assets; and (3) attracting and retaining a large enough investment community to participate in the funding opportunities. This report summarizes these challenges so that if proposed federal changes are made, stakeholders have an understanding of the possible outcomes.

The first section of this report summarizes current MLP and REIT markets and proposed rule changes. The second and third sections detail challenges to financing renewable energy assets with REITs and MLPs, respectively. Finally, the fourth section discusses possible investor responses to a renewable energy REIT or MLP.

Chapter 2 – One of the requirements for solar electricity[1] to achieve broad adoption and cost competitiveness with traditional forms of energy generation in the United States is widely accessible and low cost financial capital. The Department of Energy's SunShot goals, which would result in solar energy contributing to 14% of the nation's total electricity production by 2030, would require approximately $250 billion of solar electric generation deployment (DOE 2012). Currently, solar electric generation projects are primarily financed by institutions deploying private capital, which often require a high return on their investment. As a result, there has been a growing interest in financial policies and structures that would help migrate financing of renewable energy projects from private sources of capital to public capital markets (Mendelsohn 2012).

One investment vehicle that has been discussed by legal and financial experts in the project finance community to accomplish this is a real estate investment trust (REIT). While REITs have historically provided capital for buildings, they have recently been used to help finance other types of property, such as cellular towers and transmission lines.[2] The purpose of this study is to examine the fundamental physical, functional, and operational characteristics of a photovoltaic (PV) system in the context of the characteristics of "real" property as defined by the tax code, to help determine whether REITs can own PV systems. The solar market has treated PV as "personal" property for federal and state tax purposes. If the IRS were to classify PV as "real" property, it would likely have implications on existing market structures. This work has been reviewed and informed by real estate, tax and project finance attorneys as well as experts in PV technology and deployment, however should not be viewed as an exhaustive study of the legal and tax issues surrounding the topic.

Chapter 3 – A bill to level the playing field by giving investors in renewable energy projects access to a decades-old corporate structure with a tax advantage now available only to investors in fossil fuel-based energy projects.

At a time when the United States needs to increase domestic energy production and leaders of both political parties say they support an "all of the above" energy strategy, Congress should level the playing field and give all sources of domestic energy — renewable and non-renewable alike — a fair shot at success in the marketplace.

The federal government should not be in the business of picking winners and losers in the energy market, but for nearly 30 years, that's exactly what it has been doing with a provision in the tax code that authorizes the formation of master limited partnerships (MLPs). An MLP is a business structure that is taxed as a partnership, but whose ownership interests are traded on a market like corporate stock.

Chapter 4 – Expanded investment in clean and renewable energy resources continues to be a policy priority of the Obama Administration and an area of interest to the Congress. In recent years, the primary policy vehicle for promoting investment in renewable energy has been tax credits, particularly the renewable energy investment and production tax credits. A lack of tax liability, however, has limited the renewable energy sector's ability to fully take advantage of these and other tax benefits. The result has been an increased interest in exploring other options for promoting investment in renewable energy. One option might be to allow renewable energy entities access to the master limited partnership (MLP) business organizational form.

An MLP is a type of business structure that is taxed as a partnership, but whose ownership interests are traded on financial markets like corporate stock. Being treated as a partnership for tax purposes implies that MLP income is generally subject to only one layer of taxation in contrast to publically traded C corporations, which are subject to two layers of taxation. The ability to access equity markets in a manner similar to corporations allows MLP to obtain greater amounts of capital. Access to a greater pool of capital, when combined with the favorable partnership tax treatment, may allow MLPs to secure capital at a lower cost than similar businesses operating under a different organizational structure. The lower cost of capital, in turn, could increase investment in the renewable energy sector. Congress first established rules relating to MLPs in the 1980s. At that time, the MLP structure was limited to businesses deriving 90% of their income from primary sources, which included dividends, interest, rents, capital gains, and mining and natural resources income. Effectively, this definition allowed oil and gas extraction and transportation activities access to the MLP structure, while renewable energy resources were generally excluded. The Emergency Economic Stabilization Act of 2008 (P.L. 110-343) expanded the definition of income from qualifying sources to include transportation of certain renewable and alternative fuels, such as ethanol and biodiesel.

Provisions enacted under the American Recovery and Reinvestment Act of 2009 (ARRA; P.L. 111-5) enhanced the value of certain renewable energy tax credits for many renewable energy projects. By transforming existing tax credits into cash grants, the Section 1603 grants in lieu of tax credits program enhanced access to capital for many in the renewable energy sector. The Section 1603 grant program is scheduled to expire at the end of 2011. As Congress evaluates other policies to attract additional capital to the renewable energy sector, allowing renewable energy entities to structure as MLPs might be one option.

Extending the MLP structure to renewables could possibly attract additional capital to and stimulate investment in the renewable energy sector. There are, however, a number of potential policy concerns to consider. First, expanding access to the MLP structure could narrow the corporate tax base, which is one of the reasons access to this structure was limited in the first place. Second, if changes to the tax code allowing renewable entities to access the MLP structure are enacted alongside changes to current passive activity loss rules, there may be concerns about the possibility of renewable energy investments being used as a tax shelter. Finally, if the concern is that renewable entities are disadvantaged relative to fossil fuels currently able to use the MLP structure, one option would be to prevent other energy entities from structuring as MLPs.

In: Master Limited Partnerships …
Editor: Josiah Heslin

ISBN: 978-1-63117-012-6

Chapter 1

MASTER LIMITED PARTNERSHIPS AND REAL ESTATE INVESTMENT TRUSTS: OPPORTUNITIES AND POTENTIAL COMPLICATIONS FOR RENEWABLE ENERGY*

David Feldman and Edward Settle

EXECUTIVE SUMMARY

Master Limited Partnerships (MLPs) and Real Estate Investment Trusts (REITs) are two proposed investment vehicles that have the potential to lower the high cost of capital for renewable energy assets—a critical factor in the U.S. Department of Energy's goal for renewable energy to achieve grid-parity with traditional sources of electric generation. Due to current U.S. federal income tax laws, regulations, and administrative interpretations, REITs and MLPs cannot finance a significant portion of the cost of renewable energy assets. Efforts are underway to alter these rules by changing the definition of "real property" (REIT) and "qualified income" (MLP). However, even with rule changes, both investment vehicles have structural challenges to efficiently finance renewable energy assets. Among them are (1) effectively utilizing the

* This is an edited, reformatted and augmented version of the Office of Energy Efficiency & Renewable Energy, NREL/TP-6A20-60413, dated November 2013.

U.S. federal income tax incentives; (2) administratively structuring the investments to not be overly onerous or complicated, given the potential for pooling a relatively large amount of small assets; and (3) attracting and retaining a large enough investment community to participate in the funding opportunities. This report summarizes these challenges so that if proposed federal changes are made, stakeholders have an understanding of the possible outcomes.

The first section of this report summarizes current MLP and REIT markets and proposed rule changes. The second and third sections detail challenges to financing renewable energy assets with REITs and MLPs, respectively. Finally, the fourth section discusses possible investor responses to a renewable energy REIT or MLP.

1. Introduction to Master Limited Partnerships and Real Estate Investment Trusts

Master Limited Partnerships (MLPs) and Real Estate Investment Trusts (REITs) are investment vehicles organized under sections 7704 and 856 of the U.S. Internal Revenue Code (Code or IRC), respectively. These business structures combine advantages similar to those of a traditional publicly traded corporation with those of a private partnership. Public trading provides broad access to capital and liquidity, while a partnership-like model can avoid the double taxation that would otherwise occur at both the corporate and investor level in a public corporation.[1]

MLPs and REITS are in many ways like a mutual fund; in fact, REITs are modeled directly off of the mutual fund structure. Mutual funds are equity investments registered with the Securities and Exchange Commission (subject to various federal and state regulations) and publicly traded on market exchanges. Like mutual funds, MLPs and REITs are designed to allow a typical retail investor to make investments that the investor may not otherwise have the opportunity to make; namely, modest investments in assets (or pools of assets), such as real estate or natural gas pipelines. MLP and REIT assets have certain attractive qualities similar to debt, Treasuries, or high-quality bonds because they are designed to produce long-term, stable cash flow distributions for their passive investors.[2]

Key advantages of MLPs and REITs are highlighted as follows:

- Long-term, stable cash flows, typically through long-term, stable contracts[3]
- Little to no MLP or REIT level tax, resulting in lower overall taxes on investment
- Transferable ownership units, available to a wide pool of investors (accredited, institutional, retail, or otherwise)
- Liquidity of investment, providing for easier exit
- Public exposure, providing transparency in price
- Portfolio of assets, offering diversification benefits
- Historically higher yields to investors than the S&P 500
- Ability to raise capital for corporate needs, such as acquisitions, through public market sale of units.

These benefits have attracted considerable attention in the investor community, resulting in a lower cost of capital. Currently, approximately 172 REITs are listed on major exchanges and have a market cap of $600 billion (NAREIT 2013a). There are also approximately 120 MLPs trading on major exchanges with a market cap of $400 billion (NAPTP 2013).

However, although the two structures share many of the same benefits and are often paired together, they have different sets of governing rules.

1.1. REITs

1.1.1. Background and Current Status

REITs were established in 1960 under the *REIT Act* contained in the *Cigar Excise Tax* Extension (NAREIT 2013). The effective elimination of corporate-level income tax was designed to give the same tax treatment to investors as they would have received if they had directly invested in the underlying assets. There were two intended benefits of this action. “First, Congress believed that this method of investment would give small investors the opportunity to pool their resources in a way that would secure for them investment advantages normally available to those with larger resources. Second, Congress wanted to remove taxation as much as possible as a factor in determining the relative size of investments in stocks and securities on the one hand and real estate equities and mortgages on the other" (Kogan 2012). This was particular important at the time because of the shortage of capital in real estate, which primarily relied on “government- guaranteed money, and investments by special groups, such as insurance companies and pension trusts" (Kogan 2012).[4]

The qualifications for being a REIT are specified in Section 856 of the Code (26 USC 856). In short, to be considered a REIT, an entity must pass two tests:

- Asset test: 75% of a REIT's assets must be real property, cash, and cash items (such as receivables).[5]
- Income test: 95% of a REIT's gross income must come from approved passive income sources (such as dividends, interest, and rent) and 75% of its gross income must come from real estate.

Most REIT managers structure their investments to be well over the income and asset threshold requirements so that any unforeseen fluctuation does not disqualify the entity.[6] Due to these tests, a REIT is limited in the types of assets in which it can invest. Over the past 50 years, the variety of assets in which REITs have invested has grown (e.g., buildings, railroads, mobile homes, LED billboards, cellular towers, and electric transmission assets). However, REITs are still limited, in large part, to what the IRS considers to be real property. Figure 1 shows a timeline of a limited selection of asset classes which the IRS has interpreted in a private letter ruling (PLR) or revenue ruling (RR) to be real for purposes of REIT requirements, and the year in which the determination was made.

However, land, buildings and certain improvements to such real estate are still the most common REIT assets, as well as other assets clearly related to *bona fide* real estate such as certain qualified real estate mortgages. Though the IRS has over the years substantially expanded the list of qualified REIT activity, the focus has nonetheless remained on traditional assets.

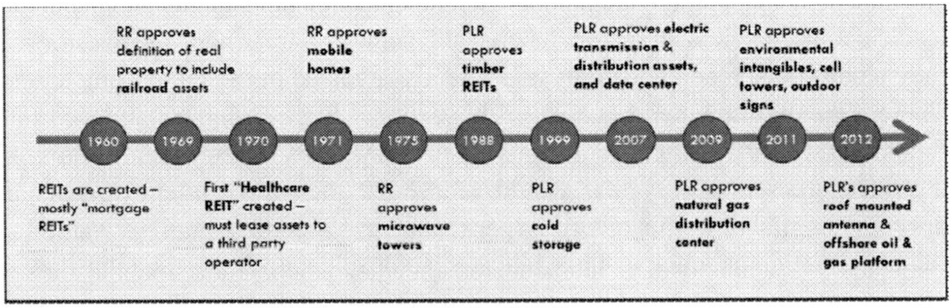

Sources: Feldman et al. (2012), and NAREIT (2013).

Figure 1. Expansion of REIT asset classes.

With respect to the portion of their assets (up to 25%) that do not need to be real property, cash, or cash items, many REITs own interests in Taxable REIT Subsidiaries (TRS). A TRS is a company, some or all of whose stock is owned directly or indirectly by a REIT, which has made a joint election with the REIT to be treated as a TRS. A TRS can engage in any business activity, including an activity that the REIT itself is prohibited from performing. In return, the TRS is required to pay corporate-level taxes, with only its after-tax profits available for distribution to its shareholders.[7]

A REIT computes its taxable income like any other corporation, except that it is also allowed to adjust its net taxable income by deducting the portion of its taxable income that it distributes to unit holders (i.e., REIT shareholders). Because REITs are required to distribute at least 90% of their taxable income to unit holders, they typically avoid paying most, if not all, taxes at the corporate level.

1.1.2. Proposed Broadening Definition of Qualifying Assets/Income

Due to asset and income limitations, REITs currently cannot own a meaningful amount of renewable energy assets. Efforts are underway to expand the definitions of qualifying assets and/or income to include renewable energy.

A renewable energy asset would qualify for REIT purposes under federal income tax rules if it was considered real property for federal income tax purposes.[8] While renewable energy technology has been interpreted as equipment (personal property) under some provisions of the Code, some technologies, such as solar,[9] share the same traits as some assets that qualify as real property under the REIT provisions.[10] For more information on this topic, refer to NREL's report, "The Technical Qualifications for Treating Photovoltaic Assets as Real Property by Real Estate Investment Trusts" (Feldman et al. 2012). The definition of real property for REIT purposes is frequently interpreted by the IRS. A company can request that the IRS make a targeted determination by asking for a PLR, in which the IRS would determine whether the company's assets qualified as real property under the REIT rules. While a PLR would only apply to that individual company, the IRS is required to treat all taxpayers consistently, and a favorable ruling could encourage broader adoption of the underlying principles by other companies who believe they have the same type of assets as those covered in the ruling. The IRS could also issue a revenue ruling, preemptively, to address whether certain assets are real property.[11] Currently, several companies are preparing requests for a PLR, are in discussions with the IRS, or have submitted a formal request to try to get renewable energy assets to qualify as real property. As of publication, the authors are unaware of any ruling that has significantly altered the tax treatment of renewable energy assets, or the likelihood that one will be issued in the future.

If the efforts to make income from renewable energy assets qualify under REIT rules are successful, the implementation of investment by these vehicles will be dependent on the language in which rule changes are drafted. While altering the definitions of qualified income under IRC Section 856 are paramount to this endeavor, other sections of the Code and the market may dictate the success of using these proposed sources of capital for renewable energy assets.

1.2. Master Limited Partnerships

1.2.1. Background and Current Status

In 1980, the top tax rate on corporate income was 46% and the top tax rate on individual income was 70%. Private partnerships were taxed at the individual level only, while investors owning shares in a public corporation were double taxed—both corporate and individual taxation.[12]

Apache Oil Company launched the first publicly-traded partnership in 1981, and the tax- advantaged structure rapidly gained popularity. The partnership was not subject to U.S. federal income tax as a corporation, so partners were only taxed at the individual level. In addition, unlike previous partnerships, Apache began trading on the established securities market with broad, attractive access to public capital, as a wider range of investors could participate in an investment opportunity that was more affordable and more liquid than a private partnership (NAPTP 2013).

Restaurants, amusement parks, the Boston Celtics, hotels, and other businesses began shifting to operate as MLPs. Between 1981 and 1987 there were over 100 MLP initial public offerings (IPO) (Fenn 2011). Fearing corporate tax avoidance and expanding erosion of tax revenue, in 1987 Congress enacted IRC Section 7704 (26 USC 7704), which treats partnerships whose interests are traded on a securities market as corporations for U.S. federal income tax purposes.[13] Subsequently, many of the real estate owning entities became REITs,[14] while other MLPs went private, were acquired, or otherwise abandoned the MLP structure.

MLPs, referred to as publicly traded partnerships in the Code, are not required to pay corporate taxes, provided that 90% or more of income generated is derived from specific sources. The general categories of qualifying sources of income are:

- Interest, dividends, and capital gains
- Rental income and capital gains from real estate[15]
- Income and capital gains from **natural resources activities**
  - Natural resources include oil, gas, petroleum products, timber, etc.
  - Natural resource activities include gathering, processing, pipelines, etc.
- Income from commodity investments
- Capital gains from sale of assets used to generate the above types of income.

The above sources can generally be divided into two categories: various forms of passive income, such as rent, dividends, and interest; and exhaustible natural resources, such as mining or transporting minerals or timber. In the 1988 amendments to IRC Section 7704, Congress included language in the definition of qualifying income intended to clarify that it does not include income and gains from renewable energy sources.[16] "Congress said that 'fishing, farming ... [and] hydroelectric, solar, wind, or nuclear power production' are not activities that deal in minerals or natural resources. Inexhaustible resources, even if natural resources, do not qualify. Examples of inexhaustible resources are soil, sod, turf, water, air and minerals from sea water. Thus, wind farms, solar power plants and hydroelectric projects are generally not suitable assets for MLPs" (Martin 2006).[17]

Currently, avoiding double taxation through reliance on the partnership provisions of the U.S. income tax code may still be attractive. As of the date of this publication, from a regulatory perspective, the U.S. corporate federal income tax rate is 35%, the tax rate on dividends is 15% to 20%, and individual tax rates range from 10% to 39.6%.

A prime example of an MLP in today's public marketplace is an energy pipeline operator that moves crude oil, natural gas, and refined petroleum products through its network of pipelines. The MLP is paid fees based on pipe capacity and volumes transported, not on the price of the underlying commodity in the pipe. Pipeline building requires significant capital, resulting in a captive market for customers and a barrier to competition. More successful MLPs are often paired with a sister corporation, which builds and then sells the qualifying source assets to the MLP, taking on the commodity and construction risk to the benefit of the MLP. As commodity demand rises, the MLP's earnings and dividends grow. The MLP can grow as well by acquiring the income from other pipelines or building new pipelines.

MLPs generally distribute their income to the unit holders on a quarterly basis[18] and are characterized as having relatively low risk with relatively high yields (5% to 10%). Distributions are usually treated as tax deferred return of owner capital, lowering the owner's tax basis while taxing gains only upon the sale of units.

As illustrated in Figure 2, more than 90% of MLP investment is currently in "natural resource" activity. Oil and gas dominate the sector, and with the rise in shale gas, the total market cap, and the oil and gas industry's share of it, should continue to grow.

In summary, the MLP structure, by avoiding double taxation via federal partnership tax rules, deferring taxes, and providing access to a broad investor base through reliance on the public securities markets, provides a lower cost of capital than a traditional public corporation structure or a private partnership, which is crucial for a capital intensive industry. Assets with relatively low returns can be built while still providing an adequate yield for investors. MLPs distribute all available cash, which is attractive to investors seeking income with little investment risk, such as pensioners.

1.2.2. Proposed Changes of "Qualifying Income" for MLPs

The renewable energy industries, particularly businesses that own and operate residential- to utility-scale solar photovoltaic assets, share many of the characteristics of the oil and gas pipeline business. Income is steady per asset deployed and is based on a long-term contract in the form of a power purchase agreement or equipment lease agreement. Income is often based off of the regional regulated utility rates for electricity. Upfront capital expense is intensive and operations and maintenance expenses are relatively low, making the cost of capital particularly important to a project's economics. The solar energy industry, like the shale gas industry, has been growing rapidly in recent years, with annual installations growing from 160 MW in 2007 to 3,341 MW in 2012 (Sherwood 2009, Sherwood 2013). Given the significant benefits associated with the MLP structure, many within the renewable energy industry are advocating for the ability to use the structure.

Presently, a limited amount of renewable energy assets are owned by publicly traded partnerships. In fact, solar energy systems can be found at most remote pipeline gathering locations, providing energy to the data and telemetry systems crucial to pipeline operations.

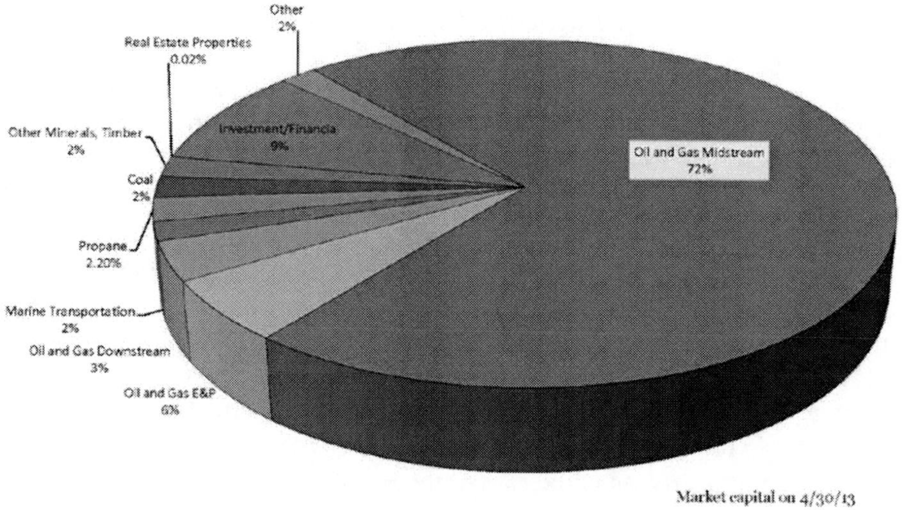

Source: NAPTP (2013).

Figure 2. MLP market cap by sector.

However, if 10% or more of the partnership's income is derived from renewable energy,[19] then it will not qualify as an MLP, and thus must pay corporate taxes.

A PLR by the Internal Revenue Service dated February 19, 2008, analyzes the issue of MLPs and renewable energy power production (IRS, PLR-200821021). The petitioner was intending an initial public offering as an MLP based on power plants fueled by various renewable energy resources. The letter refers to the Senate Report accompanying the Technical and Miscellaneous Revenue Act of 1988 that clarified that "qualifying income does not include, for example, income from … hydroelectric, solar, wind, or nuclear power production." The PLR made explicit that income derived from the sale of electrical energy generated using renewable resources not previously listed, such as geothermal energy and wood chips from timber, were not qualifying income.

An act of Congress is necessary to open up the MLP structure for broader inclusion of renewable energy assets.[20] Several proposals have been discussed (including removing MLPs from the Code altogether). However, the most prominent efforts include the bipartisan *MLP Parity Act*, first introduced in June 2012 by Senators Chris Coons (DE) and Jerry Moran (KS). The bill was later re-introduced on a bipartisan basis into the new congress in April 2013 by

Senators Chris Coons (DE), Debbie Stabenow (MI), Jerry Moran (KS), and Lisa Murkowski (AK).

As written, the proposed bill changes the definition of qualified income to include income derived from energy technologies that qualify under Sections 45 and 48 of the Code (production tax credit and investment tax credit, respectively), which are primarily renewable energy technologies, such as wind, solar, and geothermal. The proposed bill also includes income from "green" investments such as energy efficiency; grid-connected electrical energy storage; carbon capture and sequestration; and production, transportation and storage of renewable fuels and chemicals (Coons 2013).

If the Code is revised to permit renewable energy to use the MLP structure, such arrangement should not carry the perceived or apparent difficulties in reinterpreting its different sections with regards to real property and personal property, as in the case of the REIT business model (discussed in further detail in section 2). The MLP change would merely add power generation from renewable resources to the list of qualifying income activities.

If the efforts to make income from renewable energy assets qualify under REIT and MLP rules are successful, the attractiveness of these investment vehicles will be dependent on the final language of the rule changes. While altering the definitions of qualified income under IRC Sections 7704 and 856 are paramount to this endeavor, other sections of the Code, and the private sector's comprehension and comfort with these changes, will ultimately determine the success (or not) of using these proposed sources of capital for renewable energy assets.

2. Considerations if Changes Are Made to REITs

2.1. Joint Ownership

An important factor in any IRS ruling on whether solar property constitutes real property is how the asset is divided into real or personal property, if at all. It has been argued that a solar energy asset is a self-contained system which must work holistically to generate electricity. However, within this single system are many different components and materials, such as modules, inverters, conduit, and racking. In its real property PLRs, the IRS typically concludes that some, but not necessarily all, of a project's assets are real property. For example, Revenue Rule 75-424, 1975-2 C.B. 269 (IRS 1/27/12) concerns which components of a microwave transmission system are real estate assets. In the ruling, the IRS determined that the transmitting and receiving tower itself was real property. However, the transmitting, multiplex, and receiving equipment, which did not require external support to operate, were held to be assets accessory to the operation of a business and, therefore, were not real property.

If a portion of the solar property causes the REIT to fail either the gross income or asset test, then a REIT would likely not be able to solely own the solar property and maintain its REIT status. Instead, a joint ownership structure could be instituted in which the REIT owns the real property, and a non-REIT owns the personal property. While this complicates financing, potentially increasing costs that could make the process uneconomical, this approach is currently used in at least one instance.[21] For example, the company "PowerREIT" operates in the solar energy space under current IRS guidelines by owning the land underneath solar energy systems. The company, which has closed transactions in Massachusetts and California, is able to pay for a portion of the initial capital structure of a solar energy asset (thus lowering the initial outlay by the solar property owner), and generate revenue through lease payments from the land. If a larger portion of the solar energy asset were classified as real property, then it is conceivable that a REIT could contribute that additional portion of the original capital structure as well.

Certain complications may arise, particularly in the solar rooftop space, when connected assets are owned by different entities, for example, in the case of bankruptcy. This circumstance is common in the solar rooftop space due to third-party entities owning solar energy assets on other companies' or homeowners' roofs. There may be difficulties disentangling the solar asset and the roof in the event of default if they are both considered "permanent". This becomes particularly problematic for balance of plant equipment, which may serve both the solar system and the building. However, to a certain degree, this has been solved for REIT ownership by the company Hannon Armstrong. Hannon Armstrong, which raised $167 million in an initial public offering in 2013, is a REIT which primarily finances energy efficiency assets in buildings. Like plumbing and electric, energy efficiency assets are typically assimilated into the underlying building. In some instances, these buildings are owned by other entities, yet Hannon Armstrong was still able to structure these transactions to finance the ownership of the energy efficiency assets. It is believed that solar assets may also be able to find solutions to joint ownership in the event that only a portion is deemed real property.

2.2. Complications with Existing Tax Code

Designating solar assets as real property may cause unintended consequences that could significantly reduce the desired benefits of this new classification. Solar properties enjoy two prominent tax-related benefits: the investment tax credit (ITC), covered in IRC Section 48, and the five-year Modified Accelerated Cost Recovery System (MACRS) depreciation schedule, covered in IRC Sections 167 and 168. The sections of Code that govern these benefits treat solar energy property primarily as tangible personal property.

However, IRC Section 48(a)(1) and Treasury Regulation 1.48-1(c), which concerns federal investment tax credits, allows property on which a tax credit is claimed to be either "tangible personal property" or "other tangible property." If a solar asset is designated real property, then it cannot be tangible personal property, but it has been argued that it could be considered "other tangible property." In IRC Section 48, only certain types of real property are prohibited from receiving an investment tax credit, including buildings and structural components of buildings. If property serves as a "building component," then it is part of the building and cannot be used for an ITC. However, if the asset provides a "special service," it is not considered to be part of the building, and can receive the ITC while being treated as realty because its primary purpose is not building related, but energy related. The IRS has addressed "building components" for solar equipment in two separate PLRs (one involving a solar roofing membrane [PLR-200947027 {August 11, 2009}], and one involving a solar collector for a window [PLR-201043023 October 23, 2009}]). The IRS ruled in both instances that the property was not a building component, but rather performed a "special service," meaning they could receive the ITC. Despite these rulings, there is still uncertainty over the matter, and solar owners and developers may be very concerned over the potential loss of such a large tax benefit.[22]

The depreciation schedule faces similar challenges. Real property is typically depreciated over a period of 27.5 or 39 years (or not depreciated at all, in the case of land). However, the relevant section of the Code[23] states that all property eligible for the ITC,[24] including solar "equipment" (i.e., personal property), is also eligible to use the five-year MACRS depreciation schedule. If solar equipment is reclassified as real property, it may cast doubt as to which depreciation schedule is now most appropriate.

Finally, the 1603 Treasury Program, which awards a cash grant in lieu of a 30% ITC, only gives grants on equipment, not real property. Businesses must have started construction of their asset before January 1, 2012; however, grants can be given to these projects placed in service before January 1, 2017. If a change in classification were made, this may affect projects which have not received a grant yet.

Because the Code is written to implement laws from the legislative branch, and these laws are not necessarily consistent, it also does not necessarily have to be internally consistent. It may be possible to conclude, based on legislative intent and judicial interpretations, that property that qualifies as real property for one purpose (e.g., REITs) may nonetheless qualify as personal property for another purpose (e.g., depreciation). The way in which any change in classification is written will be important and may need further clarification by future rulings.

2.3. Structural Issues with Using Tax Benefits

Assuming that solar assets are eligible for using the five-year MACRS depreciation schedule and the investment tax credit, there are structural difficulties with REIT investors benefiting from such an arrangement. An advantage to forming a REIT is that the entity is only required to pay taxes on its undistributed taxable income, and because it is required to distribute at least 90% of its taxable income annually to its shareholders, and often distributes more, REITs pay little or no income taxes. Therefore, the ITC may be of little value to a REIT, particularly if solar assets are a relatively large portion of its total asset base. In addition, a REIT cannot pass along the ITC to its unit holders because it is not a pass-through entity like a partnership, whose partners are allocated shares of the partnership's income and losses (through a K-1 form); a REIT distributes dividends to its unit holders, and the unit holders receive form 1099-DIV from the REIT at the end of the year showing their individual dividend income, which they then report on their tax return. In addition, to the extent a REIT distributes its taxable income to its shareholders, a pro rata portion of any investment tax credit it would otherwise be eligible to claim is disallowed.[25]

Unit holders of a REIT *can* benefit from the depreciation of its assets. Cash distributions by REITs are taxed in different ways depending on the price one paid for ownership in the REIT and the amount that the REIT's assets have been depreciated. In fact, REITs are often thought of as tax-deferred investments because the cash distributions in initial years are not taxed due to depreciation expenses offsetting income at the REIT level, resulting in a non-taxable return of capital at the investor level. However, because the five-year depreciation schedule is so accelerated, the depreciation expense is often greater than a solar asset's income, which generates a tax loss. Unit holders of a REIT are unable to take advantage of this loss by sheltering other income; they can only shelter the distribution they receive. Therefore, because these losses must be used in future years, much of the time value of the depreciation is lost. If solar assets were a significant portion of a REIT's asset base, the REIT would either lose much of the benefit of the ITC and accelerated depreciation,[26] or need to structure a financial arrangement with an alternative equity source (tax equity provider), which would limit the amount of capital it could contribute.[27]

Another option which companies are currently exploring is purchasing portfolios of solar energy assets after their tax benefits are fully utilized (and the projects are allowed to be sold without triggering the recapture rules in the Code). In doing this, a REIT would avoid the issue of how to use the tax benefits at the REIT level (since they were used by the prior owner), and thus have a portfolio of assets whose primary benefit is long-term stable cash flow (like other REIT asset classes). If REIT qualifications do change for solar, this may be one of the first strategies to be implemented. Conversely, because of the solar market's rapid growth, not many solar projects are five years or older. A REIT could purchase a solar asset that received a Treasury grant (as the same recapture rules do not apply); however, the REIT acquisition must not have been planned as a method of circumventing the issue.

2.4. Prohibited Transaction Tax on Electricity Sales

A REIT cannot be used to own an operating business, and therefore, if it sells inventory, it is subject to a 100% tax on profits from the sale. Electricity is typically classified as inventory for companies that own generating assets and use them to generate electricity for sale (as opposed to leasing them to someone else). If electricity were to qualify as inventory for purposes of a solar asset, then its sale by a REIT would attract a tax equal to 100% of the REIT's net income from those sales.[28] In order for a solar REIT to generate revenues through power purchase agreements without it being a prohibited transaction, there would need to be a determination by the IRS that electricity generated by the REIT is not inventory, or Congress would need to pass legislation incorporating an exception into the rules (Congress passed this exception for the sale of timber in *The Food Conservation and Energy Act of*

2008). An alternative would be to structure all transactions using a lease in lieu of a PPA. However, this practice is currently only used in the distributed solar space, and is likely to be problematic for a project contracted for a utility.

2.5. Complications with Third Party Ownership

If solar assets were confirmed to be real property by the IRS, this could create uncertainty for current financing structures that use either leases or power purchase agreements (PPAs). Third- party businesses have gained significant market share in the United States, driving a considerable amount of distributed PV demand. For example, as noted in Figure 3, approximately 90% of residential systems installed in Arizona during 2012, and through the first half of 2013, were third-party owned.

Tax treatment of a lease as a true lease and not a capital lease or financed sale of the asset must meet certain requirements established primarily through case law. In 2001, the IRS reiterated its views on the factors supporting treatment of a lease as a true lease.[29] Among these factors are issues relating to permanence: 1) the lease term cannot be more than 80% of the life of an asset; 2) the owner must have a claim for the residual value of the asset; and 3) the lessor must have the ability to offer the services of the asset to someone else other than the lessee at the end of the lease. If it does not pass these tests, the leasing of the asset could be treated as a sale of the asset to the lessee. A solar asset's treatment as real property, and thus inherently permanent, might cause problems for a transaction passing the third, and possibly the second, test.

2.6. Other Complications

Solar assets currently receive property tax exemptions in many states. These laws often treat solar assets as personal property. This does not necessarily create a conflict with treating solar assets as real property for REIT purposes; federal and state laws often classify the same assets in different categories. Further, the REIT rules expressly state that local law definitions do not control for purposes of determining the definition of real property under the REIT rules.[30] For example, many local laws classify property, such as machinery, as fixtures while federal law defines it as personal property. However, it should be identified as a potential complication, particularly in states and localities whose laws follow federal laws.

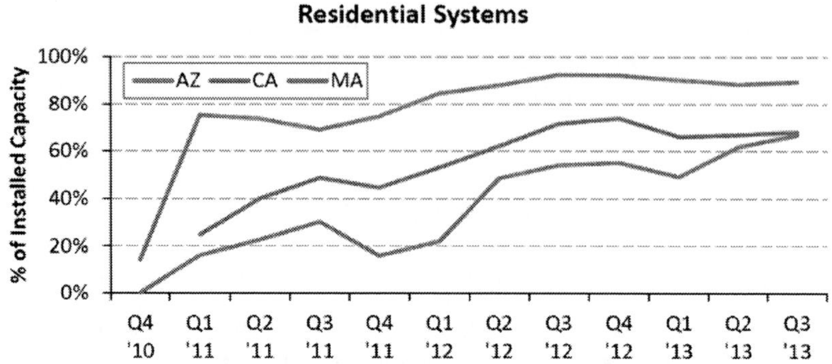

Source: Arizona Goes Solar (2013); CSI (2013); DOER (2013).

Figure 3. Percentage of residential PV capacity installed as third-party systems in select states.

Finally, if solar assets are classified as real property, an investor in solar energy projects may incur tax on gains from the sale of such property under the Foreign Investors Real Property Tax Act ("FIRPTA"), as codified in Sections 897 and 1445 of the Code. In general, foreign persons are taxed in the United States only on investment income from U.S. sources and income that is connected with the conduct of a trade or business in the United States by the foreign person. In contrast, gain from the sale of property is not generally subject to tax in the United States. However, FIRPTA changes the latter rule in the case of gains from the sale of U.S. real property and treats such gains as if they had been derived by the foreign person from a U.S. trade or business. The foreign seller of such property is subject to U.S. tax on the net gain from the sale, and the buyer is required to withhold 10% of the purchase price of the property and remit it to the IRS as a "prepayment" of the seller's U.S. tax liability.[31] Thus, if solar is classified as real property, foreign investment in projects will also be subject to these rules, which may dampen investment.

3. Considerations if Changes Are Made to MLPs

3.1. Loss Limitations for Investors

Investments in partnerships are much more complicated to track than investments in corporations. A partner's investment in a partnership is tracked through two accounts. The first is the capital account, which represents the partner's share of the equity contributed to the partnership. The second account calculates a partner's outside basis, or the amount of a partner's investment in his or her interest in the partnership. This amount determines how much gain (or loss) the partner would incur upon a sale of that interest.

A partner's capital account begins as the sum of the cash and the market value of any property contributed by the partner to the partnership. A partner's outside basis begins as the sum of the cash and the partner's tax basis in property contributed by the partner to the partnership.[32] Both amounts then increase by the amount of partnership income allocated to the partner and decrease by the amount of cash distributed and losses allocated to the partner. If the partnership has nonrecourse debt, then the partner's share of this debt is added to his or her outside basis, and the partner's share of any repayment of that debt is subtracted from his or her outside basis.[33] These rules are particularly relevant to solar projects because the use of their accelerated depreciation schedule, and the potential use of debt, which can cause significant losses and low capital account and outside basis amounts in the first several years of a project.

If a partnership incurs an operating loss, partners who are *individuals* or *closely-held corporations* are subject to certain *loss limitation* rules in the Code that can prevent these investors from fully utilizing this loss.

- **Capital basis limitation:** a partner cannot claim a loss to the extent it exceeds the amount in the partner's capital account. When the capital account is reduced to zero, it means that for tax purposes, the partner has no capital invested in the partnership. The Code's partnership rules do not allow a partner to claim losses beyond the partner's equity investment in the partnership.[34]
- **At-risk limitation:** Assuming basis is greater than zero, the partner's capital must be at risk or the loss cannot be claimed. Capital may not be at risk if there are, for example, non-recourse loans, or loan guarantees that protect the investor's capital. Like the capital basis limitation, this rule is designed so that a partner cannot receive benefits without risk in its investment. If the only capital at risk, for tax purposes, is non-recourse debt (in which the debt is secured only by the project, not by the partners), then the partners are not entitled to benefit from the losses.
- **Passive loss limitation:** Passive activity income can only be offset by passive activity losses. Passive activity is characterized as a partner who does not materially participate in the business. Other examples of passive income include rental income or investment income. Passive

losses cannot be deducted from a partner's earned income, such as a salary. To the extent passive activity income exceeds passive activities losses in a particular year, the excess loss is carried forward to future years. For publicly traded partnerships, *a loss in one MLP cannot be used to offset income from a separate MLP, or any other investments generating passive income.*

The loss limitations are designed to prevent investors from using losses from investments, in which they are passive or in which they have no risk, to offset an investor's income from other activities. These rules also apply to tax credits as well, meaning a partner of an MLP is limited to using tax credits it receives from the MLP to offset taxes owed on income it receives from investment in the same MLP. Currently, two of the main benefits of solar energy generation ownership are the tax losses associated with the accelerated depreciation schedule and the 30% ITC. Due to these loss limitation rules, it would be difficult for an individual or closely held corporate partner in an MLP to benefit from the accelerated depreciation and ITC without significant loss of the time-value of these benefits.[35]

3.2. Mixed Portfolio Considerations

One method for avoiding the loss limitation rules to an MLP is to combine solar assets with other qualifying assets into one MLP. Losses and tax credits generated by a solar asset could be used to offset income and taxes owed from other assets, such as a natural gas pipeline, to the extent they are realized under the rules set forth above. Today's MLPs are ordinarily focused on a specific niche business, and business focus is often viewed by the investment community as a characteristic of a well-managed business. However, MLPs are constantly evaluating their ability to grow the asset base and cash distributions. This may include, for example, addition of businesses that are directly related to the core business going beyond mere pipeline operations and adding various assets for gathering, processing, and shipping natural gas, natural gas liquids, and refined oil and gas products. Occasionally, the partnership may expand into an indirectly related business, such as fuel ethanol storage tanks, an asset category which was added as an MLP qualifying income source in *The Energy Improvement and Extension Active of 2008*.

All else being equal, renewable energy businesses, such as third-party solar energy companies, may be attractive to those MLPs willing to evaluate them strictly as a growth opportunity that will add to the asset base, enhance depreciation charges, increase cash distributions to investors, and ultimately increase the distributions from the partnership. A diversified portfolio may achieve greater results through pairing certain types of assets (e.g., natural gas combined with hydroelectric power or solar combined with wind). A diversified portfolio may require more analysis of timing, tax avoidance, and cash flow benefits.

Market research might be able to gauge on a hypothetical basis whether existing MLPs or their investors would be interested in diversifying into renewable energy investments. Likewise, additional market research could help understand whether investors would be willing to invest in an MLP that is solely focused on renewable energy. Such studies could be useful for structuring transactions and compiling the prospectus in advance of an IPO, but ultimately, until several renewable energy MLPs are available to the market, the depth of investor interest cannot reliably be predicted. With that said, a similar business model has recently been deployed in the public capital markets by NRG. In July 2013, NRG Yield successfully raised over $400 million through an IPO, with a diversified portfolio of contracted renewable energy projects and conventional generation and thermal infrastructure assets in the United States.[36] NRG is in a unique position because it is one of the largest renewable energy and traditional energy asset owners in the United States. However, based on the IPO, public markets appear willing to invest in a mixed- generation portfolio of assets, including solar.

3.3. Partnership Tax Management Challenges

Financial reporting, tax accounting, corporate governance, and other compliance responsibilities of publicly traded entities can be quite complicated; even more so for MLPs. There are nuances that must be addressed differently from those entities taxed as corporations, beginning with the initial public offering and continuing into "normal operations." Prior to an IPO, the various business structures and assets must be aligned for consistency in a pass-through environment. After the IPO, varying classes of equity such as common, preferred, and general partnership units must be tracked and managed. Continued growth and integration is also imperative to increase cash distributions to unit holders, demanding roll-ups, drop-downs, roll-outs, acquisitions, and divestitures.

Managing a solar MLP has the potential to be even more complicated. Depreciation charges are one of the very attractive features in an MLP, sheltering income from tax. A partnership owning a solar or wind project is usually in a net tax loss position due to depreciation until the fourth year after the project is placed in service. Any excess depreciation is passed through to partners. They can carry it forward for up to 20 years. The excess depreciation is usually enough to shelter the income from a project for the first nine years after the project is placed in service (assuming they do not use the excess depreciation in earlier years). Assets like buildings and pipelines do not truly wear out, according to the schedules set forth by the IRS, but nonetheless, those long depreciation time frames allow for a long, steady return of capital to unit holders. Similarly, high-quality solar energy property is guaranteed by manufacturers to produce at more than 80% of its nameplate capacity after 25 years, while MACRS depreciation on the property is only five years.

In the case of a third-party residential solar MLP, which manages and owns equipment on thousands of homes, each system would have a distinct tax fixed asset depreciation schedule, based on the date the equipment was placed in service. Managing all of these assets has the potential to complicate an already difficult accounting procedure. The responsibilities of a publicly traded partnership are well understood today by those active in the business, just as the responsibilities of managing a renewable energy company are well understood; the complexity will be in the marriage of these businesses.

3.4. Challenges to Attracting Investors to MLPs

Expanding the definition of “qualified income” to include income from solar assets would give the solar industry another vehicle in which to access public markets, thus potentially expanding the amount of capital able to be deployed in the industry, and ideally lowering the cost of this capital. However, current MLP laws also make it challenging for some investors as well, and their limitations would also apply for a solar MLP. First, unlike many forms of public investment, MLPs are partnerships, which have different accounting and reporting methods. Whereas a shareholder in a publicly-traded corporation might receive a 1099 stating either the gain or loss on shares for a given year, a unit holder in a partnership receives a K-1 form. The K- 1 form shows the investor’s share of each item of partnership income, gain, loss, deductions, and credits, and must be reported on the investor's tax return. Additionally, for income from MLPs operating in multiple states, the investor will receive a state K-1 and may be required to file a state income tax return in each of those states with reported income. Further, MLPs must complete the partnership tax returns before the K-1 forms can be completed and sent to investors. Finally, K-1 amendments are frequent, and require amendments to individual taxpayer returns. Thus, many MLP investors find it necessary to request an extension for income tax filing. These unique investment characteristics specific to partnerships may disincentivize investment by average retail investors.

Because income from MLPs is often tax deferred, there is likely to be less inducement for investment from non-taxed accounts such as IRAs, 401(k)s, and other qualified retirement accounts. Further, if a tax-exempt entity invests in an MLP, the Code requires that they be treated as engaged in an "unrelated" trade or business and must pay taxes on the income as "unrelated business taxable income" (UBTI).[37] All UBTI above $1,000 is taxed at the highest rate for a trust, currently 39.6% (NAPTP, 2013). A significant portion of U.S. investment is from tax-exempt entities, such as IRAs, universities, and pension funds.

Foreign investors also have more complicated tax filing requirements when investing in an MLP. Foreign investors are required to file tax returns reporting their share of an MLP's income and must withhold taxes at the highest applicable effective tax rate on MLP quarterly distributions. The foreign investors receive credit back on excess withholdings only after they have filed a U.S. federal tax return.

The tax consequences, filings, and withholdings of MLP investment may discourage investment from less sophisticated investors, foreign investors, and non-profit organizations. While this would not be unique to a solar MLP, it should be recognized that allowing solar to qualify as "good income" for MLP purposes does not necessarily expand its potential investors to all U.S. public capital market participants.

3.5. Steady, Predictable Revenue

MLPs distribute cash on a quarterly basis, and MLPs generally predict, without guarantee, the amount of the quarterly distribution. Investors choosing the MLP vehicle often anticipate a steady distribution of cash from their investment.

Production of electricity from solar assets is reasonably predictable and based on the equipment choice, angle, and positioning of the modules related to the path of the sun and other environmental and operational factors. However, production from solar is not steady throughout the year.[38] If the customer agreements are based on actual production during each quarter, as opposed to levelized production over the year, the cash available for distribution each quarter will vary considerably.

4. Comparison of Solar MLP/REIT Investment to Traditional MLP/REIT Investment

Investment decisions are traditionally determined by an individual investor's specific balance of desired risk and return. U.S. Treasury securities are globally liquid and considered riskless because they are backed by the full faith and credit of the U.S. government. The remainder of the risk spectrum generally involves a range from money market funds and high-quality bonds to high-risk equities. MLPs and REITs have returned relatively high yields for their volatility compared to other investment choices over the last decade. Risk and return are typically considered highly correlated. In other words, the riskier an investment, the higher its expected return must be to attract investors. As compared to other investment classes in Figure 4, MLPs and REITs have higher returns than other investments with similar volatility (i.e. standard deviation, or risk), and have lower volatility than investments with similar returns. These attractive qualities of MLP and REIT securities has caused demand from investors and produced a relatively large and stable investment industry.

One of the advantages to expanding the eligibility of REITs and MLPs to include solar equipment is that solar development would have access to an entire industry of lawyers, financiers, and investors with the understanding, and experience, to deploy billions of dollars in capital efficiently and effectively through REITs and MLPs. In theory, by tapping into this knowledge-base and infrastructure, the solar industry will be able to more easily and more cheaply deploy assets.

Certain characteristics of the solar industry differ from assets which have traditionally been financed in MLP or REIT vehicles. Most significantly, unfamiliarity with the asset class will require investors to be educated about the nuances of the industry. The U.S. solar industry has grown rapidly in recent years; 70% of total installed capacity came on-line in 2011-2012 alone (Sherwood 2013). Because it is a much newer asset class than the housing market, for example, there may be more perceived investment uncertainty, and thus riskiness, making solar REITs or MLPs more expensive than their more traditional counterparts. However, over the past 30 years, potential investors have been educated on the risk and return profiles of new asset classes, and as a result MLP and REIT securities have been issued in enterprises such as ethanol storage, cemeteries, fertilizer, and carbon dioxide, among others. Perceived riskiness of solar assets should decline over time through additional information provided to investors, such as the credit quality of PPA (or lease) countersigners, expected lifetime of asset, and the production and volatility of cash-flows from solar investments. Further, new production forecasting methods are in development to better estimate a solar asset's electricity production, which will also reduce uncertainty.

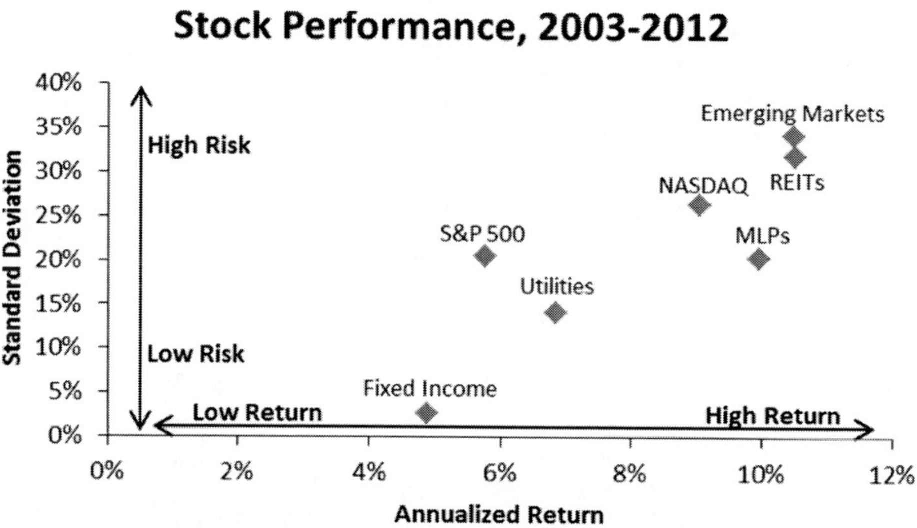

Source: finance.yahoo.com. Annual returns and standard deviation of annual returns based on adjusted close for stocks and indices with proxies of asset classes as follows: emerging markets (iShares MSCI Emerging Markets [pricing only available from 2004-12]); fixed income (iShares Core Total US Bond Market ETF); MLPs (Alerian MLP Index); REITs (iShares Dow Jones US Real Estate); utilities (PHLX Utility Sector [pricing only available from 2004-12]).

Note: much of volatility in REIT return is due to the housing market crisis in 2007-08.

Figure 4. Returns and standard deviation across asset classes.

It can be argued that the underlying risk of owning a solar asset is equivalent to or lower than many traditional MLP and REIT asset classes. The revenue stream of a midstream oil and gas MLP has potential volatility due to the current and future price of natural gas. A certain portion of its income is derived from the volume of natural gas processed and transported; if there is an increase in natural gas price, demand can fall leading to a reduction in income for the midstream MLP. If natural gas prices rise in the future, this could also impact a midstream MLP's ability to renew or replace existing agreements. Furthermore, a midstream MLP faces the risk of government regulatory action, disaster (such as hazardous material spill or loss of life, man-made or otherwise), and higher than expected operations and maintenance costs. Finally, midstream MLPs have contracting risks due to the underlying creditworthiness of their customers and their ability to negotiate long-term contracts. As shown in Table 1, some natural gas pipeline contracts may be structured with durations longer than 10 years; however, other MLP sectors have much shorter average underlying contract lengths.

Appendix A highlights some of the risks given by the MLP Tallgrass Energy Partners in its recent initial public offering, as may be compared to those of a solar investment.

Table 1. Midstream MLP Business Profiles

Cash Flow Stability	Type of Business	Contract Length	Revenue Type	Type of Customers
Very High	Natural gas pipelines	**10+ years**	Rental fee/"Ship-or-pay"	Gas distributors, utilities, producers, marketers and others
	Crude oil pipelines	**5-10+ years**	Rental fee/Volume	Refiners, producers, financials
	Storage	**3-5 years**	Rental fee/Volume	Utilities, marketers, financials
	Refined prod. pipelines	**1-5 years**	Rental fee/Volume	Refiners, marketers

Cash Flow Stability	Type of Business	Contract Length	Revenue Type	Type of Customers
	NGL pipelines	**1-5 years**	Rental fee/Volume	Petrochemical plants, producers
	Gathering	**Ranging from month-to-month to life of lease dedications**	Rental fee/Volume	Producers
	Fractionation	**Typically short-term contracts but trending more long-term**	Fee-based/ "Frac-or-pay"	Producers
	Rail	**1-5 years**	Fee-based/Distance	Refiners, producers
	Terminals	**1-3 years**	Volume/Ancillary services	Refiners, financials
	Processing	**Month-to-month to life of lease dedications**	Fee-per-car/Distance	Producers
	Marine Shipping	**1-3 years**	Fee-based/Indexed charter rates	Refiners, petrochemical companies, integrateds, marketers
Very Low	E&P	**-**	Market rates/Hedging	Midstream operators

Source: Maresca et al. (2013).

A REIT which owns commercial property also has inherent investment risk. The underlying value of commercial properties is subject to change due to changes in population, employment, or ownership trends. This has a direct effect on a REIT's ability to maintain and renew long-term leases and manage the creditworthiness of their customers. REITs are also subject to costs beyond their control, such as regulatory developments that affect their ability to dispossess or evict an occupant or increase rent. These costs can also include increases in real estate taxes, fees, insurance premiums, litigation costs, compliance costs, and any unforeseen environmental or structural issue with a building or property.

A solar MLP or REIT would have a similar, or perhaps lower, risk profile to their traditional counterparts. Like a natural gas pipeline, a solar energy asset has a degree of revenue volatility due to underlying annual and seasonal weather variability. The ability of a solar REIT to replace or renew existing agreements is similarly dependent on future competitive pricing. There is also the risk of future regulatory action, though the chance of a disaster is far lower for a solar facility than a natural gas pipeline or other midstream operations. Increased operations and maintenance costs are also less, due to fewer moving parts in a solar asset compared to a building or pipeline. With an expected life exceeding 30 years, a solar energy system typically has a long contract length compared to other industries using MLPs or REITs, with power purchase agreements (or leases) ranging from ten to twenty-five years. Finally, the underlying revenue stream of an asset is dependent upon the creditworthiness of its customer. Large solar energy systems often have agreements with regulated utilities; 99% of the public utilities rated by Moody's and Fitch were investment grade (UWEB 2012). Transactions in the residential solar energy space are made with homeowners, with varying degrees of credit. However, most financing agreements, thus far, have required the homeowner to possess a minimum credit score of 680 to 700 (Marciano 2013).

In summary, while there are risks associated with solar energy systems, they are reasonably similar and, in several aspects, less risky than many industries currently financed through MLP and REIT structures. In addition, perceived investment risk within the solar industry should decrease over time with increased information available to investors on asset and investment performance.

CONCLUSION

MLPs and REITs have the potential to lower renewable energy assets' high cost of capital because of their ability to access public markets and the avoidance of double taxation. In order for MLPs and REITs to own a meaningful amount of solar energy assets, the IRS would have to change its definition of what qualifies as real property (for REITs) or legislation must be passed by Congress to change the definition of "qualifying sources of income" for MLPs.

The details of these changes, if they were to happen, could play an important part in determining the success with which MLPs and REITs are adopted to own solar energy assets, due to the current complexities within the Code and the investment profile of solar energy assets. Financing solar equipment under the REIT structure may face challenges in maximizing the benefits of the accelerated depreciation schedule and the investment tax credit (though this is expected to become less of an issue in the future as installed prices—per watt—fall, and the tax credits expire). There may also be conflicting dynamics with current leasing practices within the solar market, foreign investment, and state property tax laws, if solar equipment is determined to be real property. MLPs also face complications due to loss limitation rules; the accounting and legal complexities of structuring MLPs; and the added tax procedures for retail investors, non-profits, and foreign investors.

If these challenges are overcome, investors may be convinced to make investments in solar MLPs or REITs because the underlying risk of owning a solar asset is equivalent to or lower than many traditional MLP and REIT asset classes, while offering the same predictable cash flows. Expanding the eligibility of REITs and MLPs to include solar equipment would also give the solar industry access to an entire industry of lawyers, financiers, and investors with the understanding, and experience, to deploy billions of dollars in capital efficiently and effectively through REITs and MLPs. In theory, by tapping into this knowledge base and infrastructure, the solar industry will be able to more easily and more cheaply deploy assets, helping the U.S. Department of Energy's goal of solar achieving grid-parity with traditional sources of electric generation (DOE 2011).

References

Arizona Goes Solar. (2013). *Arizona Public Service and Salt River Project installation data*. http://arizonagoessolar.org. Accessed October 2, 2013.

California Solar Initiative (CSI). (2013). "California Solar Statistics." http://www.californiasolarstatistics.ca.gov/. Accessed September 30, 2013.

Coons, Christopher. (2013). "*The Master Limited Partnerships Parity Act.*" Accessed August 8, 2013: http://www.coons.senate.gov/issues/master-limited-partnerships-parity-act.

DOE. (2011). "*DOE Pursues SunShot Initiative to Achieve Cost Competitive Solar Energy by 2020.*" Accessed September 25, 2013: http://www1.eere.energy.gov/solar/sunshot/news_detail.html?news_id=16701.

Feldman, D., Mendelsohn, M. & Coughlin, J. (2012). *The Technical Qualifications for Treating Photovoltaic Assets as Real Property by Real Estate Investment Trusts.* NREL/TP-6A20-55396. Golden, CO: National Renewable Energy Laboratory. Accessed March 27, 2013: http://www.nrel.gov/docs/fy12osti/55396.pdf.

Fenn, T. (2011). "MLP IPO Trends." Midstream Business.

Internal Revenue Service. (October 24, 2011). IRS PLR 201204006. Accessed 2013: http://www.irs.gov/pub/irs-wd/1204006.pdf.

Internal Revenue Service. (August 11, 2009). IRS PLR 200947027. Accessed 2013: http://www.irs.gov/pub/irs-wd/0947027.pdf.

Internal Revenue Service. (October 23, 2009). IRS PLR 201043023. Accessed 2013: http://www.irs.gov/pub/irs-wd/1043023.pdf.

Internal Revenue Service. (July 30, 2012). IRS PLR 201323016. Accessed 2013: http://www.irs.gov/pub/irs-wd/1323016.pdf.

Internal Revenue Service. (February 19, 2008). IRS PLR 200821021. Accessed 2013: http://www.irs.gov/pub/irs-wd/0821021.pdf.

Internal Revenue Service. (2013). "IRS Publication 925, Passive Activity and At-Risk Rules." Accessed September 25, 2013: http://www.irs.gov/uac/Publication-925,-Passive-Activity-and- At-Risk-Rules.

Kogan, K. (2012). "*Can REITs Invest in Renewable Energy Projects*?" Bloomberg BNA: Daily Tax Report.

Marciano, J. (May 17, 2013). "The March Toward Securitzation and SolarCity's $500 Million Goldman Transaction." *Chadbourne & Parke Project Finance News*. Accessed August 14, 2013: http://www.pfnewswire.com/2013/05/the-march-toward-securitzation-and.html.

Maresca, S., Kad, R., Sheikh, S. & Lasky, B. (2013). *Midstream Energy MLPs Primer 3.0.* Morgan Stanley Research.

Martin, K. (2006). "Master Limited Partnerships." *Chadbourne & Parke Project Finance News: March 2006.*

Martin, K. (2007). "A New Structure for MLP Roll Ups." *Chadbourne & Parke Project Finance News: January 2007.*

Martin, K. (2013). "New Financing Strategies." *Chadbourne & Parke Project Finance News: August 2013.*

Massachusetts Department of Energy Resources (DOER). (2013). "RPS Solar Carve-Out Qualified Units." http://www.mass.gov/eea/docs/doer/rps-aps/solar-carve-out- units.xlsx. Accessed September 30, 2013.

National Association of Publicly Traded Partnerships (NAPTP). (2013). "Master Limited Partnerships 101: Understanding MLPs." Accessed August 8, 2013: http://www.naptp.org/documentlinks/Investor_Relations/MLP_101.pdf.

National Association of Publicly Traded Partnerships (NAPTP). (2013a). "REIT Industry Timeline." Accessed August 8, 2013: http://www.reit.com/timeline/timeline.php.

National Association of Real Estate Investment Trusts (NAREIT). (2013). "REIT Industry Timeline." Accessed August 8, 2013: http://www.reit.com/DataAndResearch/US-REIT-Industry-MarketCap.aspx.

National Association of Real Estate Investment Trusts (NAREIT). (2013a). "US REIT Industry Equity Market Cap." Accessed August 8, 2013: http://www.reit.com/DataAndResearch/US-REIT-Industry-MarketCap.aspx.

Oppenheimer Funds. (2011). "Master Limited Partnership Primer: Understanding an Emerging Asset Class."

Revenue Procedure 2001-28 - Leveraged Leases. (2001). Accessed September 25, 2013: http://www.novoco.com/energy/resource_files/irs_guidance/rulings/proc_01-28.pdf.

Sherwood, L. (2009). *U.S. Solar Market Trends 2008*. Interstate Renewable Energy Council (IREC). Accessed 2013: http://irecusa.org/irec-programs/publications-reports.

Sherwood, L. (2013). *U.S. Solar Market Trends 2012*. Interstate Renewable Energy Council (IREC). Accessed 2013: http://irecusa.org/irec-programs/publications-reports.

Eugene Water & Electric Board (UWEB). (2012). "Memorandum to Commissioners, Subject: Credit Ratings History and Context." Accessed August 16, 2013: http://www.eweb.org/public/commissioners/meetings/2012/120904/Corr_CreditRatingHistoryan dContext.pdf.

U.S. Congress. 26 United States Code (USC) 856 - Definition of real estate investment trust.

U.S. Congress. 26 USC 46 – Amount of credit.

U.S. Congress. 26 USC 50 – Other special rules.

U.S. Congress. 26 USC 48 - Energy Credit.

U.S. Congress. 26 USC 7704 - Certain publicly traded partnerships treated as corporations.

U.S. Congress. 26 USC 168 - Accelerated cost recovery system.

U.S. Congress. 26 USC 857(b)(6)(A) - Taxation of real estate investment trusts and their beneficiaries.

U.S. Congress. (1988). House Report 100-795, at 400.

U.S. Congress. (1987). *Revenue Act of 1987 (P.L. 100-203)*, §10211(a). HR 3545. 100th Congress, first session. H 100-495.
U.S. Congress. (1988). Technical and Miscellaneous *Revenue Act of 1988 (P.L. 100-647)*, §2004(f)(4). HR 4333. 100th Congress, second session. H 100-1104.
U.S. Congress. (1988). Senate Report 100-445, at 424.
U.S. Congress. Treasury Regulation 1.856-3(d) Definitions – Real Property.

APPENDIX A.

The table below summarizes several risk factors recently published in an initial public offering prospectus for a natural gas transportation and storage services partnership, Tallgrass Energy Partners, LP. It indicates whether there may be a corollary risk factor for a hypothetical solar energy partnership.

Table A-1. Comparison of Midstream Energy Partnership and Corollary Solar Energy Partnership Risk Factors

Midstream Energy Partnership Risk Factor	Corollary Solar Energy Partnership Risk Factor
Level of firm transportation and storage capacity sold and the volume of natural gas we transport, store and process.	Solar energy sold to a subscribed customer would be based on sunshine, not market or consumer demand, once the initial contract is executed.
Level of production of oil and natural gas and the resultant market prices of natural gas and NGLs.	Depending on sunshine, the level of production is reasonably established at the initial contract. Market prices influence new contracts only.
Actual and anticipated future prices of natural gas impacting ability to renew and replace agreements.	Actual and anticipated future prices of electricity may impact the ability to extend the power purchase or lease agreement with a host.
Regulatory action affecting supply, demand, rates, operating costs, operating flexibility, or method of contracting.	Regulatory action could affect rates, operating costs, or method of contracting.
Effect of seasonal variations in temperature on the amount of natural gas transported, stored, processed, and treated.	Temperature effects would be minimal; however, the amount of sunshine each quarter may differ materially from the longer term average forecast.

Midstream Energy Partnership Risk Factor	**Corollary Solar Energy Partnership Risk Factor**
Competition from other midstream energy companies.	Competition would be material for signing up new solar customers, but would not impact current sold properties and the underlying revenues.
Customer creditworthiness.	Customer creditworthiness.
Operating and maintenance costs.	Although the solar industry is less mature, the technology has far fewer moving parts and labor requirements than the natural gas industry.
Damages to equipment and properties from natural disasters and acts of terrorism.	Solar properties may face similar natural disaster risks, and vandalism in place of terrorism.
Outages at processing facilities, requiring extended and complex repairs.	Solar facilities may experience outages, with the inverter being a weaker link. Repair timing and complexity could be comparable to a home HVAC system or hot water heater.
Leaks or accidental releases of hazardous materials into the environment.	Not a risk factor for solar.
Cost of acquisitions affecting actual cash available for distributions, debt service requirements, etc.	Similar risk factor for solar.
Extent to which customers are willing to contract on a long-term basis.	Solar faces a similar risk as the technology matures, customer profiles change, competition continues to increase, and power prices fluctuate.
Poor acquisition choice and assumptions.	Similar but lower risk anticipated given low technology complexity, established records for operation, no environmental liability, etc.
Price and availability of other forms of energy.	Solar energy rates are approaching grid parity as the technology adoption expands rapidly. Should another technology accomplish a similar path, it could present a risk to solar energy.
Increased costs to explore for, develop, produce, gather, process, and transport natural gas.	Solar enjoys a relatively steady profile for energy production. After the first 25 years of operation, solar modules continue to produce at more than 80% of original capacity.
Nature and extent of, and changes in, governmental regulation such as greenhouse gas legislation.	Solar energy is likely to receive favorable treatment under greenhouse gas legislation.
Hazards that result in personal injury and loss of life.	Solar energy requires fewer employees and they are exposed to materially fewer risks than those employees working in the natural gas industry.

End Notes

[1] While an MLP is literally a partnership for federal income tax purposes, a REIT is not. However, special rules applicable to REITs (discussed later) provide them with a tax effect similar, though not identical, to a partnership, while not actually being a partnership and not using partnership income tax law.

[2] See IRS Publication 925 for a definition of "passive."

[3] Investor return from a REIT is, in large part, structured to come from dividends (like coupons on a bond). However, unlike debt, a REIT is not required to return the REIT owner's original investment; the investor may effectively receive this through sale of the asset. In other words, debt issued at 6% must pay back principal in addition to the coupons; a REIT structure's predictable cash flows provide a 6% yield, and the original investment is paid back when the investor sells their units.

[4] This is not dissimilar to current capital constraints in the renewable energy space, which rely on government tax credits and tax equity investors.

[5] The asset test must be met on a quarterly basis. Both parts of the income test must be met on an annual basis.

[6] Subject to certain limited exceptions, a REIT that fails to meet either test is disqualified from electing the REIT tax treatment for five years.

[7] Unlike REITs, TRSs do not have to distribute their profits.

[8] See Treas. Reg. § 1.856-3(d), which defines "Real Property." Note, the definition of real property is *not* provided for in the tax code itself, but rather in the regulations.

[9] For purposes of this paper, "solar" refers to photovoltaic generation devices, as opposed to concentrated solar power, solar water heating, or solar energy component manufacturing.

[10] The sticking point of the current IRS position is that machinery cannot comprise more than 15% of the assets or income of a REIT unless the machinery serves only the building or other structure with which it is associated (thus, for example, solar panels that are able to delivery electricity into the grid at any point in time would not be REIT real property). The IRS holds to its position on machinery even where the machinery at issue is in and of itself an inherently permanent structure (such as a wind-driven generator or a utility-size solar facility). Many renewable energy facilities meet the requirement for inherent permanence: the facilities are attached to real property (e.g., land, buildings, parking canopies) and will ordinarily remain affixed indefinitely. The IRS has accepted the REIT-ability of many classes of non-building structures that conduct or store commodities such as electric and gas distribution assets because those assets are inherently permanent. At the same time the IRS has denied REIT treatment for similarly permanent assets that produce a commodity, such as renewable energy facilities. The distinction that the IRS draws between conduit/storage assets and productive assets is a distinction first imposed when the original REIT regulations were adopted in 1962. Since the distinction is not driven by the statute or by Congressional intent, the IRS can, by regulation alone, enable renewable energy REITs that invest in inherently permanent facilities.

[11] A Revenue Rulings represents the IRS's interpretation of how the law applies to a particular set of facts. Unlike a PLR, which technically applies only to the taxpayer that requested it, a revenue ruling applies to all taxpayers with the same facts.

[12] In a "traditional corporate structure," corporate profits are subject to corporate federal income taxes, and shareholders of the corporation are (usually) subject to personal federal income taxes when they receive dividends or distributions of those profits. Thus, there is a double layer of taxation, or double taxation.

[13] *Revenue Act of 1987 (P.L. 100-203)*, §10211(a). Congress made IRC Section 7704 prospective only and allowed master limited partnerships that were in existence prior to the effective date of this provision to retain partnership status as long as they do not substantially expand or terminate their business activities.

[14]The REIT structure is often viewed as more attractive to investors from a tax compliance perspective. Jeff Eckel, CEO of the REIT Hannon Armstrong states, "A REIT is friendlier to investors than an MLP. Investors get simple 1099 forms at year end with the amount of their dividends. They do not have to fuss with complicated K-1 forms reporting cash distributions, allocations of various kinds of income, capital accounts, outside bases and the like (Chadbourne, 2013)." In addition, investors in MLPs are considered engaged in business in the states where the MLP does business and have to file state tax returns.

[15] There is an interesting intersection between the REIT and MLP provisions of the Code: income from real property—rents, interest, and gains—are good income under the MLP provisions because the MLP provisions expressly adopt the REIT definitions of rent, income, and gains.

[16] Technical and Miscellaneous Revenue Act of 1988 (P.L. 100-647), § 2004(f)(4). See also House Report 100-795, at 400 (1988), Senate Report 100-445, at 424 (1988).

[17] Efforts have been made to structure MLPs that avoid qualified income issues by forming a series of partnerships which issue dividends from non-qualified sources of income, and that are partially owned by the MLP. Because the MLP receives the cash flow as a dividend, which is passive, instead of non-qualified income, it has changed the nature of the cash flow. These arrangements are very hard to structure and come with a series of complications. See Martin (2007) for more details.

[18] Two upstream MLPs, Linn and Vanguard, have adopted monthly distributions.

[19] In other words, if 10% or more of the partnership's income is derived from renewable energy, then less than 90% would come from "qualifying sources."

[20] The exception to this is "rental income and capital gains from real estate," which would be qualifying income under the presumption that solar assets were to qualify as real property. Again, the likelihood of this happening is unknown.

[21] See, e.g., PLR 201323016 (holding land upon which a solar system was located was eligible for REIT ownership).

[22] However, this issue will be diminished if the tax credit reverts from 30% to 10% as scheduled in 2017.

[23] 168(e)(3)(B)(vi)(I)

[24] Section 48(a)(3)

[25] See IRC Section 50(d)(1), which applies rules "similar to" IRC Section 46(e) effective prior to the date of enactment of the Revenue Reconciliation Act of 1990. Prior IRC Section 46(e)(1)(B) and (2)(B) limited the REIT to claiming a share of the ITC equal to the same pro rata share of taxable income that the REIT retained. The remaining portion of the ITC was disallowed. Further, if the REIT pays taxes on a smaller share of its income in any of the next four years, then the ITC it claimed may be subject to partial recapture.

[26] REITs can carry unused losses forward to reduce REIT taxable income in later years, though would lose some of its "time value."

[27] Certain financial arrangements (such as a partnership, or inverted lease) with a tax equity provider may also face complications due to a REIT's tax classification. For example, if a tax equity provider entered a partnership with a REIT for the purpose of owning solar assets, the REIT may inhibit the tax equity provider from receiving the full ITC and benefits from the tax losses.

[28] See section 857(b)(6)(A)

[29] See Rev. Proc. 2001-28.

[30] See Treas. Reg. § 1.856-3(d).

[31] Exceptions from this rule are available for certain investors who own interests in publicly traded REITs, or REITs in which foreign investors own in the aggregate less than 50% of the interests. However, applying these rules in practice is often difficult.

[32] If a partner purchases his or her interest in the partnership from another person, outside basis starts as the amount paid by the partner for that interest.

[33] Partnership-level nonrecourse debt does not affect the partners' capital accounts.

[34] Note, however, that there is an exception to this rule to the extent the losses are funded with nonrecourse debt allocated to the partner. This rule allows a partner to take deductions for losses that are funded with the partner's share of nonrecourse debt (e.g., nonrecourse debt is used to purchase property that generates depreciation deductions) even if those losses exceed at some point the amount in the partner's capital account (this amount is referred to as minimum gain). In return, as that debt is repaid, the partner is required to recognize gain equal to his or her share of the repaid debt (this amount is referred to as minimum gains chargeback income). The minimum gains chargeback income partially offsets the advantage of taking the depreciation expense.

[35] Another complication with a partner in an MLP using the ITC is that if it sells its interest in the MLP during the first five years after a project is put in service, it will have a partial recapture of the ITC claimed on the project.

[36] A close review of the IPO prospectus for NRG Yield discloses that NRG Yield has developed a tax strategy to manage its taxable income in such a manner that, for a lengthy period of time, the bulk of its distributions to shareholders are treated as non-taxable returns of capital. Retaining the ability to distribute "return of dividend" non- taxable distributions will be difficult for NRG Yield given the manner in which accelerated depreciation and other beneficial tax attributes are effectively reversed in the "earnings and profits" account, an account which must be zeroed out ("cleansed") in order to make return-of-capital distributions to shareholders. NRG Yield will have difficulty continuing to produce non-taxable distributions and it will be even tougher for others to replicate.

[37] Further complications may arise as tax exempt ownership typically inhibits a company from claiming an ITC.

[38] It should be noted that many businesses are seasonal, including natural gas pipelines. This may support the idea of bundling different asset classes into one MLP or REIT to potentially avoid peaks and troughs.

In: Master Limited Partnerships ...
Editor: Josiah Heslin

ISBN: 978-1-63117-012-6

Chapter 2

THE TECHNICAL QUALIFICATIONS FOR TREATING PHOTOVOLTAIC ASSETS AS REAL PROPERTY BY REAL ESTATE INVESTMENT TRUSTS (REITS)*

David Feldman, Michael Mendelsohn and Jason Coughlin

CONTEXT

One of the requirements for solar electricity[1] to achieve broad adoption and cost competitiveness with traditional forms of energy generation in the United States is widely accessible and low cost financial capital. The Department of Energy's SunShot goals, which would result in solar energy contributing to 14% of the nation's total electricity production by 2030, would require approximately $250 billion of solar electric generation deployment (DOE 2012). Currently, solar electric generation projects are primarily financed by institutions deploying private capital, which often require a high return on their investment. As a result, there has been a growing interest in financial policies and structures that would help migrate financing of

* This is an edited, reformatted and augmented version of a National Renewable Energy Laboratory publication, Technical Report NREL/TP-6A20-55396, dated June 2012.

renewable energy projects from private sources of capital to public capital markets (Mendelsohn 2012).

One investment vehicle that has been discussed by legal and financial experts in the project finance community to accomplish this is a real estate investment trust (REIT). While REITs have historically provided capital for buildings, they have recently been used to help finance other types of property, such as cellular towers and transmission lines.[2] The purpose of this study is to examine the fundamental physical, functional, and operational characteristics of a photovoltaic (PV) system in the context of the characteristics of "real" property as defined by the tax code, to help determine whether REITs can own PV systems. The solar market has treated PV as "personal" property for federal and state tax purposes. If the IRS were to classify PV as "real" property, it would likely have implications on existing market structures. This work has been reviewed and informed by real estate, tax and project finance attorneys as well as experts in PV technology and deployment, however should not be viewed as an exhaustive study of the legal and tax issues surrounding the topic.

Inherent to answering this question is whether PV systems possess three fundamental properties: permanence, passivity, and being integrated as a system. These three properties are associated with real property as follows: The Internal Revenue Service (IRS) stipulates that if a system is *inherently permanent,* then it is realty,[3] and if it is an *accessory to the operation of a business,* it is not. *Inherently permanent* and *an accessory to the operation of a business* are mutually exclusive terms. Therefore, to establish whether a piece of property should be considered realty, it must be determined which term best applies based on the facts and circumstances of the particular situation. *Passivity* is one characteristic that helps determine whether an asset is an accessory to the operation of a business, and thus non-real property. It is also important to determine what pieces of a PV system should be characterized as real property. If a PV system is *integrated*, one could make an argument that all of it should be classified in the same way. Therefore, establishing whether a PV system is inherently permanent and passive will help inform whether it is real, and establishing whether it is integrated as a system will establish whether all of it should be treated in the same way. Further research should be performed to explore possible outcomes of treating PV systems as real property, and whether or not it would be practical for REITS to own them.

INTRODUCTION TO REITS

The REIT structure was designed to bundle and securitize real estate assets in order to attract capital from a wide range of sources. A REIT is a tax designation status that eliminates most corporate taxes so long as the REIT distributes 90% of its taxable income to investors.[4] The REIT structure has several attractive features compared to traditional sources of financing for PV. Many of these have the potential to lower the cost of capital for projects, increase the pool of potential financiers, and allow a large number of locations well cited for solar to be more easily developed.

Key Benefits

Some key benefits of REITs include:

- A REIT typically owns a portfolio of assets, providing diversification benefits by spreading risk over a large number of projects.
- REIT ownership, structured as transferrable shares, provides access to a wider range of investor classes, such as pension funds and foreign and retail investors.
- A REIT can be listed and traded on public exchanges, thus making it a more liquid investment.
- REITs currently own a large amount of commercial real estate that could host solar assets.

Internal Revenue Code Section 856 outlines the definitional requirements of a REIT (CFR 856- 1). In short, to be considered a REIT, an entity must pass both an:

- Income test[5]: 95% of a REIT's income must come from approved sources (such as rent)
- Asset test: 75% of its assets must be real property.

There are only two types of property according to the IRS, real and personal, and the terms are mutually exclusive. If a solar asset is designated to be personal property, it would be limited to being 25% or less of a REIT's assets and might limit the percent of income a REIT receives from the solar asset.[6] The following sections will explore the fundamental characteristics of PV systems in relation to the properties of permanence, passivity, and being integrated as a system.

PERMANENCE

Real property must have the quality of *inherent* permanence. Section 1.856-3(d) of the Income Tax Regulations, "provides that 'real property' includes land or improvements thereon, such as buildings or other *inherently permanent* structures thereon (including items which are structural components of such buildings or structures)" ((CFR 856-3)). The IRS has attempted to clarify whether an asset is "inherently permanent" with the following questions: (1) "(I)s the property capable of being moved, and has it in fact been moved? (2) Is the property designed or constructed to remain permanently in place? (3) Are there circumstances that tend to show the expected or intended length of fixation, that is, are there circumstances that show the property may or will have to be moved? (4) How substantial a job is removal of the property, and how time consuming is it? (5) How much damage will the property sustain upon its removal? (6) What is the manner of affixation of the property to the land?" (Rev Rul 80-151). Examples of assets that have passed this test include: commercial buildings, HVAC systems, cellular towers, LED billboards, and railroads.

While solar assets can and have been moved without damaging the property, particularly in ballasted rooftop systems,[7] they are designed and constructed to remain in place permanently, or at least for the useful life of the asset (which can exceed 25 years). A solar system is uniquely integrated into each building or property. System size, panel orientation and tilt, capacity of the inverter, length of wiring runs, and a host of other decisions are made based on the characteristics of the site. PV panels are typically mounted in a manner that allows for replacement or removal, but this is principally for maintenance, and when performed represents a small portion of the full system. In addition, due to concerns of theft, systems are increasingly being designed to make it difficult to extract the panels.

In addition, given the high upfront cost, recouping an investment in a PV system is usually predicated on the long-term revenue streams produced by the asset while incurring only modest costs over that same timeframe. If a business were to move a PV system, either located on a roof or ground-mounted, it would only retain the value of the physical assets. The National

Renewable Energy Laboratory (NREL) estimated the cost of hardware for installing a system to be between 50%–70% of the total cost of a system in 2010 (Goodrich et al. 2012). The business would also incur the cost of removing the system and either selling the assets (which are worth significantly less due to their inability to receive an investment tax credit or local incentive[8]) or paying the cost of building the system elsewhere. Thus, the net resale value of a system after it is installed is considerably less than the cost of the installed hardware.

For this reason, PV systems tend not be put on old roofs or moved to new locations once installed.

Historically, modules had been the majority of PV system costs; now that the price of modules has fallen dramatically, the majority of the value of a system is no longer in the modules, further reducing any economic incentive to remove the energy-generating system.

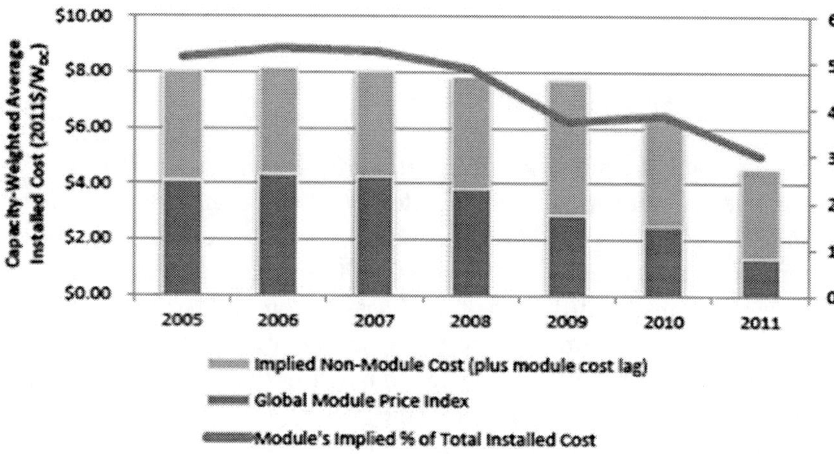

Sources: Barbose et al. 2011; SEIA and GTM Research 2012; U.S. Bureau of Labor Statistics (for inflation adjustment) 2012.

Figure 1. PV module price compared to total U.S. system costs.

In Figure 1, the PV module's implied percent of total installed system cost decreases from 51% in 2005 to 30% in 2011. Current module prices are below the 2011 value indicated, many being sold at or below $1/W. Therefore, although one could remove the panels and sell them, most of the value is not in the module.

Power purchase agreements (PPAs) and lease agreements between building/land owners and third-party solar asset owners include end-of-term removal options.

However, in the vast majority of cases the intent is not to decommission the system but rather to renew the contract or provide the host with a buyout option for its fair market value at contract termination.[9] Those arrangements are structured in this way for tax purposes that do not change the intended or preferred outcome.[10] The IRS has made similar judgments in instances with cellular towers and billboards on buildings (IRS 7/22/11; IRS 10/28/11; IRS 1/27/12). These assets could be moved, but they are not intended to be moved.

Therefore, while a PV installation can be physically moved from one location to another without causing significant damage, the system design, installation, associated contracts, permits, and project economics are all predicated on the asset remaining in a single location for the entirety of its useful life.

PASSIVITY

The IRS has determined on several specific instances that if an asset is passive in nature—such as a billboard, a tower, or a transmission line—then it is not an accessory to the operation of a business, which viewed in conjunction with its inherent permanence, makes it realty. The reason passivity of an asset is relevant is because it differentiates that asset from a "machine," which is cited as an example of an "accessory to the operation of a business."[11] In Private Letter Ruling (PLR) 147229-06 (March 13, 2007), a case involving a REIT that owned transmission systems, the IRS cites a previous ruling that a railroad track, but not the rail cars, represents real property. In that ruling, the IRS states, "similar to the tracks and other railroad components described in Rev. Rul. 69-94, the System is a passive conduit that allows (electricity) created by a generation source to flow through the system to end-users. The System itself does not include any machinery or equipment that creates or generates (electricity)... Based upon the information submitted and representations made, we conclude that the System is an inherently permanent structure that is not an accessory to the operation of a business" (IRS 6/22/07). The following will discuss three ways in which a solar system can be considered passive: how it functions; the characteristics of operating it; and whether it is involved in the underlying business of the user.[12]

How a PV System Functions

Historically, traditional energy property, such as a coal-fired plant, has not been viewed as realty due to the active nature of its processes producing electricity. A solar system operates like all energy-generating technologies in that it performs the activity of generating electricity. It has a fuel source, sunlight, which it converts into electricity. Energy is never made, it is only converted. However, a PV system is unique in that it does not require any moving parts.[13] Energy from other electric-generating assets is *required to be converted into mechanical energy (with moving parts) before it is turned into electrical energy*. Wind turbines and hydro-electric plants convert wind or water energy into mechanical energy and then into electric energy. Concentrating solar power (CSP) systems, coal plants, nuclear power plants, and natural gas facilities convert thermal energy into mechanical energy and in turn convert that into electric energy. Only a PV system does not turn energy into mechanical energy. Rather, PV cells convert radiation energy directly into electrical energy.

In addition, unlike other electric-generating assets, a PV module produces direct current (DC) electricity, which must be turned into alternating current (AC) electricity to be used by most electricity systems. The conversion is done by an inverter, which is composed of electrical components and a transformer. It may have moving parts (e.g., a fan for cooling), but its primary function is performed with no mechanical activity. An inverter, which regularly monitors the electric load to which it is connected, is an essential part of most PV systems. If the inverter cannot detect an electric grid, power does not flow. Inverters are most comparable to transformers, which are part of a transmission system. While PLRs cannot be used or cited as a precedent, it should be noted for reference purposes that in PLR-147229-06 (March 13, 2007) the IRS ruled that a REIT could own a transmission asset because it was real property (IRS 6/22/07).

Characteristics of Operating a PV System

A PV system is operated differently than most electric-generating assets in that it usually does not require an operator to be on-site every day. Like many energy-generation facilities, operating it has become automated but requires oversight, management, and maintenance. One metric that has been quantified to evaluate how much effort is necessary to maintain an energy-generating facility is its operations and maintenance (O&M) costs (which includes various labor, replacement and other costs). O&M is often perceived as relatively inexpensive for PV systems compared to other forms of power generation. While this is true of PV on a capacity basis ($/kW installed), it is not so on an energy basis ($/kWh).

Table 1. Operations and Maintenance Costs across Electric-Generating Technologies

	Combined O&M ($/kWh)	Rank (per kWh)	Combined O&M ($/kW)	Rank (per kW)
Hydro	$0.0042	1	$32	3
Natural Gas, Advanced Combined				
Cycle (CC)	$0.0050	2	$38	4
Natural Gas, CC	$0.0052	3	$40	5
Wind, Onshore	$0.0072	4	$28	2
Coal	$0.0081	5	$61	6
Natural Gas, Advanced Combustion				
Turbine	$0.0087	6	$70	8
Natural Gas, Conventional Combustion				
Turbine	$0.0094	7	$75	9
Natural Gas, Distributed Generation	$0.0094	8	$76	10
Natural Gas, Adv. CC	$0.0103	9	$78	11
Nuclear	$0.0131	10	$103	13
Photovoltaic	*$0.0134*	*11*	*$26*	*1*
Coal, Integrated Gasification CC	$0.0146	12	$109	14
Coal, Carbon Capture	$0.0180	13	$134	15
Biopower	$0.0206	14	$150	16
Wind, Offshore	$0.0226	15	$87	12
Solar Thermal	$0.0233	16	$63	7
Natural Gas, Fuel Cell	$0.0454	17	$346	17
Biopower, Landfill	$0.0590	18	$429	18

Source: Energy Information Agency, Annual Energy Outlook (AEO). O&M costs from AEO 2011; capacity factors from AEO 2009.

In Table 1, PV has the lowest O&M costs on a capacity basis. However, because PV has a low capacity factor, on a per-kilowatt-hour basis, the technology falls in the middle of the group as it relates to O&M costs.

The Underlying Business of the User

PV systems differ from many other forms of electric generation because they are designed modularly and can either be built large enough to generate as much electricity as a typical power plant or just enough to supply part of a building's electrical needs. PV systems can be sited on a building owner's property or roof or on a large parcel of land. The PV system can also be the property of the underlying building/land owner (host-owned) or a third party can own the system (third-party owned).

As noted above, in determining whether something is an accessory to the operation of a business, the IRS cites a machine as an example. Presumably, the machine's function is directly related to the business in which it is owned, such as a bottling machine at a beverage manufacturing plant.

This would be in contrast to the building, which provides the place for the business to function. In the case of host-owned PV, as long as the business of the system owner is not related to selling electricity, the PV system functions more like a building than a bottling machine. If the owner is an electric utility that is selling the electricity produced by the system, then the PV system's function of producing energy would be directly related to the underlying business.

Third-party owners typically contract PPAs or leases to generate revenue from their PPA systems. These contracts are either with the site host or an electric utility. In a PPA, the electricity produced by a PV system is sold for the life of the contract. In a lease, the lessee makes payments for the right to use the PV system for the life of the contract. In both instances, there is often an "operator" that monitors the systems and performs maintenance when necessary. If a third-party owner uses a PPA to generate revenue, its business is selling electricity produced by the PV system. In this instance, it would be difficult to argue that the PV system's function of generating electricity is not actively involved in the business. If, however, a third-party owner uses a lease[14] to generate revenue, it would be easier to argue that its business is leasing property and not selling electricity.

Passivity Summary

A PV system functions, is operated, and participates in the operation of a business in ways that are both similar and different than other electric-generating assets. The IRS will have to determine which qualities it sees as most important in determining whether a PV asset is passive in nature and thus not an accessory to the operation of a business.

INTEGRATED AS A SYSTEM

In certain instances, the IRS has concluded that assets must be separated into different classifications: some as real property and the rest as personal property. PLRs concerning LED billboards [IRS PLR-111324-11 (July 19, 2011) (IRS 10/28/11) and IRS PLR-125828-11 (October 24, 2011) (IRS 1/27/12)] and cellular towers [IRS PLR-130186-10 (April 6, 2011) (IRS 7/22/11)] were able to separate out the assets into two categories: inherently permanent structures, such as the tower or billboard structure; and property that was an accessory to the operation of a business, such as the non-permanent signs or cellular equipment.

Revenue Rule 75-424, 1975-2 C.B. 269 (IRS 1/27/12), concerns whether various components of a microwave transmission system are real estate assets. In the ruling, the IRS determined that the transmitting and receiving tower itself was real property.

However, the transmitting, multiplex, and receiving equipment, which did not require external support to operate, were accessories to the operation of a business and, therefore, were not real property.

A PV system, in many instances, also has a mounting structure that is separate and distinct from the rest of the equipment. However, the performance of a PV system would deteriorate dramatically without it.[15] With regards to the remainder of the system, removing any of the components of value (e.g., modules, inverters, or wiring) disrupts the ability of the system to function. More so, if the solar asset is not connected to the grid or a source of energy load, electricity will not flow.

A solar system's inability to function unless all of its parts are connected and working properly creates a strong argument for it being viewed as an integrated system.

A single asset, such as a PV module or inverter, cannot perform its function separately; therefore, the system as a whole should be viewed as either inherently permanent property or property accessory to the operation of a business.

Conclusion

Based on this initial examination, it would appear that PV systems have many of the qualities associated with inherently permanent assets. Furthermore, while a PV system is similar to all other electric-generation assets in many respects, it is distinct in that it does not require energy to be converted first into mechanical energy in order to then produce electricity. The inverter, which converts the DC electricity produced by the solar modules to AC electricity, is the most active component of an installation (along with the tracking system if one is included). However, an inverter is comparable to transmission assets in many respects, and the IRS has concluded in certain instances that transmission assets are real property. While a PV system does not usually require direct management on a daily basis, it does require oversight, monitoring, and maintenance, which has a cost in-line with most electric-generating assets. In addition, a PV system also may or may not be actively involved in the business that owns the system. Finally, the majority of the assets that make up a PV system would appear to meet the "integrated as a system" characteristic given the limited viability of these individual components to function independent of one another. Whether or not the IRS determines that when taken together, these qualities of a PV system characterize it as *inherently permanent* or an *accessory to the operation of a business* is beyond the scope of this analysis. The intent of this study is to provide a technical perspective to help inform the decision-making process.

If a PV system were deemed to be real property, there would still be potential obstacles to REITs owning PV assets. Among these are the ability of a REIT to fully monetize the investment tax credit as well as take advantage of depreciation and the potential differences between how state and federal authorities treat the asset (currently many states treat PV as personal property to exempt it from property taxes). In addition, the reclassification of PV as realty has the potential to disrupt current financing practices in the solar market today.

References

Barbose, G.; Darghouth, N.; Wiser, R.; Seel, J. (September 2011). Tracking the Sun IV: An Historical Summary of the Installed Cost of Photovoltaics in the United States from 1998 to 2010. LBNL-5047E. Berkeley, CA:

Lawrence Berkeley National Laboratory. http://eetd.lbl.gov/ea/ems/reports/lbnl-5047e.pdf.

Code of Federal Regulations (CFR) – Title 26: Internal Revenue: 1.856-1 Definition of real estate investment trust. Accessed 6/12/12. http://www.taxalmanac.org/index.php/Treasury Regulations, Subchapter A, Sec. 1.856-1.

CFR – Title 26: Internal Revenue: 1.856-3 Definitions. Accessed 6/12/12. http://www.taxalmanac.org/index.php/Treasury Regulations, Subchapter A, Sec. 1.856-3.

DOE (Department of Energy). (February 2012). "Solar Industry Financial Issues and Opportunities," Chapter 8. SunShot Vision Study. http://www1.eere.energy.gov/solar/sunshot/vision study.html.

Energy Information Agency (EIA), Annual Energy Outlook (AEO). (2011). Electricity Market Module Assumptions, page 97. Accessed 6/12/12. http://www.eia.gov/forecasts/aeo/assumptions/pdf/electricity.pdf

EIA, Assumptions to AEO 2009, Report # DOE/EIA-0554, March 2009.

Goodrich, A.; James, T.; Woodhouse, M. (February 2012). Residential, Commercial, and Utility-Scale Photovoltaic (PV) System Prices in the United States: Current Drivers and Cost-Reduction Opportunities. NREL/TP-6A20-53347. Golden, CO: National Renewable Energy Laboratory. http://www.nrel.gov/docs/fy12osti/53347.pdf.

IRS PLR-147229-06, March 13, 2007. Released June 22, 2007. http://www.irs.gov/pub/irswd/0725015.pdf.

IRS PLR-130186-10, April 06 2011. Released July 22, 2011. http://www.irs.gov/pub/irswd/1129007.pdf

IRS PLR-111324-11, July 19 2011. Released October 28, 2011. http://www.irs.gov/pub/irswd/1143011.pdf

IRS PLR-125828-11, October 24 2011. Released January 27, 2012. http://www.irs.gov/pub/irswd/1204006.pdf.

IRS Revenue Ruling (Rev Rul) 80-151, 1980-1 C.B. 7. Cited in IRS PLR-125828-11, October 24 2011. Released January 27, 2012. http://www.irs.gov/pub/irs-wd/1204006.pdf.

Mendelsohn, M. (2012). "Will Solar Projects Need Tax Equity in the Future? Yes, but Baby Steps Toward Securitization Improve the Situation." NREL. Accessed June 11, 2012. https://financere.nrel.gov/ finance/content/solar-PV-photovoltaics-value-tax-credits-equityaccele rated-depreciation-securitization.

SEIA & GTM Research. (March 2012). "U.S. Solar Market Insight Report: Q4 2011 & 2011 Year-In-Review."

Treasury Regulations, Subchapter A, Section 1.263A-8(c)(4). Accessed 6/12/12. http://www.taxalmanac.org/index.php/Treasury Regulations, Subchapter A, Sec. 1.263A-8.

U.S. Bureau of Labor Statistics (BLS): CPI-U. Accessed 6/12/12. ftp://ftp.bls.gov/pub/special.requests/cpi/cpiai.txt.

End Notes

[1] This paper will focus its attention on photovoltaic (PV) energy systems. However, many of the same conclusions can be applied to concentrated solar power (CSP) or solar water heating (SWH) systems.

[2] REITs can provide capital in the form of debt or equity. A REIT that provides mostly debt is typically referred to as a "mortgage REIT"; one that provides mostly equity is referred to as an "equity REIT." This nomenclature is for descriptive purposes only, as they remain the same classification of investment vehicle.

[3] Realty, real estate, and real property are all used interchangeably.

[4] A REIT can deduct its distributions against income for tax purposes, which ideally lowers a REIT's net income to close to zero. However, a REIT still has to pay taxes on the remaining net income.

[5] Technically there are two "income" tests; 75% of their income must come from certain approved sources, and 95% must come from a broader set of sources.

[6] Most REIT managers structure their investments to be well over the 95% of income and 75% of asset requirements so that any unforeseen fluctuation does not disqualify the entity.

[7] A ballasted rooftop system does not physically attach the system to the roof. Rather, a weight, such as cinderblock, is placed on the racking system so that the system does not move. This differs from modules, which are physically bolted to the roof, or poles being driven into the ground for a ground-mounted system.

[8] A person or business can only receive an investment tax credit or local incentive on new solar equipment. Moreover, many local solar programs require that an incentive be returned if a system is removed in the first 5–20 years.

[9] PPAs often have early buyout clauses between six years and the end of the transaction in addition to end-of-term buyout options.

[10] Lease accounting follows certain procedures to ensure that the agreement is treated as a lease and not a sale of the asset (PPAs typically follow the same procedures). The IRS established safe harbor guidelines to help determine whether something should be classified as an operating lease or sale. Among these are the following guidelines relating to permanence: (1) the contract length cannot be more than 80% of the life of an asset; (2) the owner must have a claim for the residual value of the asset; and (3) the lessor must have the ability to offer the services of the asset to someone else. If it does not pass these tests, the leasing of the asset could be treated as a sale of the asset. If the asset is sold, then the lessor would have no right to the depreciation or investment tax credit.

[11] The exception to this is a machine that is a structural component of a building. "In the case, however, of a building or inherently permanent structure that includes property in the nature of machinery as a structural component, the property in the nature of machinery is real property" [Treas. Reg. 1.263A-8(c)(4)]. The term "structural component" of a building is

not defined, however, can be interpreted to be a fundamental portion of a structure that is embedded and not easily removed, such as a roof or a window. While most traditional PV assets would not be characterized in this way, more advanced PV technologies are now being incorporated into windows, facades, or roofs.

[12] This paper does not attempt to resolve which form of passivity is most pertinent to the determination of realty but rather to characterize the properties of a PV system so that others can make an informed decision. In addition, while PV technology is contrasted to all other forms of electric generation, this is not to say that other renewable energy (RE) generation assets are also not passive in these respects (e.g., like PV, the fuel source for all RE has not been collected through industrialized processes and transported to the site of the electric generating asset).

[13] Though not required, many utility-scale PV systems have moving parts if they use actively tracking structures, which adjust the tilt of the panels based on time of day and wind conditions. This process is performed to produce more energy, per panel, or to prevent wind damage. PV system inverters sometimes have moving parts as well to cool the equipment, though many use passive cooling techniques, depending on the climate.

[14] The property can either be leased to the end-user or to a third-party operator, which in turn sells the electricity to an end-user. Historically, there have been lease provisions in other industries that entitled the lessor to a percentage of revenue generated from the asset.

[15] However, the racking system represents a small percentage of the total cost of the system; therefore, going to the trouble of creating a REIT just for racking assets would not appear to be a worthwhile endeavor.

In: Master Limited Partnerships ...
Editor: Josiah Heslin
ISBN: 978-1-63117-012-6

Chapter 3

MASTER LIMITED PARTNERSHIPS PARITY ACT*

U.S. Senator Chris Coons

A bill to level the playing field by giving investors in renewable energy projects access to a decades-old corporate structure with a tax advantage now available only to investors in fossil fuel-based energy projects.

At a time when the United States needs to increase domestic energy production and leaders of both political parties say they support an "all of the above" energy strategy, Congress should level the playing field and give all sources of domestic energy — renewable and non-renewable alike — a fair shot at success in the marketplace.

The federal government should not be in the business of picking winners and losers in the energy market, but for nearly 30 years, that's exactly what it has been doing with a provision in the tax code that authorizes the formation of master limited partnerships (MLPs). An MLP is a business structure that is taxed as a partnership, but whose ownership interests are traded on a market like corporate stock.

By statute, MLPs have only been available to investors in energy portfolios for oil, natural gas, coal extraction, and pipeline projects. These projects get access to capital at a lower cost and are more liquid than traditional financing approaches to energy projects, making them highly effective at attracting private investment. Investors in renewable energy projects, however, have been explicitly

* This white paper was made publicly available, April 24, 2013.

prevented from forming MLPs, starving a growing portion of America's domestic energy sector of the capital it needs to build and grow.

The *Master Limited Partnerships Parity Act* is a straightforward, powerful tweak to the federal tax code that could unleash significant private capital into the energy market.

The legislation, which is just over 600 words long, would level the playing field between traditional and new energy businesses by helping energy-generation and transmission companies form master limited partnerships, which combine the funding advantages of corporations and the tax advantages of partnerships.

By allowing additional forms of energy development to access this market tool, we can go beyond political rhetoric and start delivering an all-of-the-above energy strategy.

> MLPs combine the funding flexibility of a corporation with the tax benefits of a partnership, creatively stimulating private investment in energy projects.

The federal government should not be in the business of picking winners and losers in the energy market, but for nearly 30 years, that's exactly what it has been doing with a provision in the tax code that authorizes the formation of master limited partnerships (MLPs). An MLP is a business structure that is taxed as a partnership, but whose ownership interests are traded on a market like corporate stock.

By statute, MLPs have only been available to investors in energy portfolios for oil, natural gas, coal extraction, and pipeline projects. These projects get access to capital at a lower cost and are more liquid than traditional financing approaches to energy projects, making them highly effective at attracting private investment. Investors in renewable energy projects, however, have been explicitly prevented from forming MLPs, starving a growing portion of America's domestic energy sector of the capital it needs to build and grow.

The *Master Limited Partnerships Parity Act* is a straightforward, powerful tweak to the federal tax code that could unleash significant private capital into the energy market.

The legislation, which is just over 600 words long, would level the playing field between traditional and new energy businesses by helping energy-generation and transmission companies form master limited partnerships, which combine the funding advantages of corporations and the tax advantages of partnerships.

By allowing additional forms of energy development to access this market tool, we can go beyond political rhetoric and start delivering an all-of-the-above energy strategy.

HOW MASTER LIMITED PARTNERSHIPS WORK

An MLP is a business structure that is taxed as a partnership, but whose ownership interests are traded on a market like corporate stock. Whereas profit from publicly traded C corporations is taxed at both the corporate level and the shareholder level, income from MLPs is taxed only at the shareholder level because it is treated as a partnership for tax purposes.

An MLP consists of limited partners (investors) and general partners (managers). The limited partners — who can number in the thousands — provide capital and receive quarterly required distributions generally equivalent to shareholder dividends in a C-corporation. They play no role in the operation of the MLP, while the general partners manage the MLP's daily operations. General partners can take the form of another company or a group of individuals, typically holding a 2 percent ownership stake.

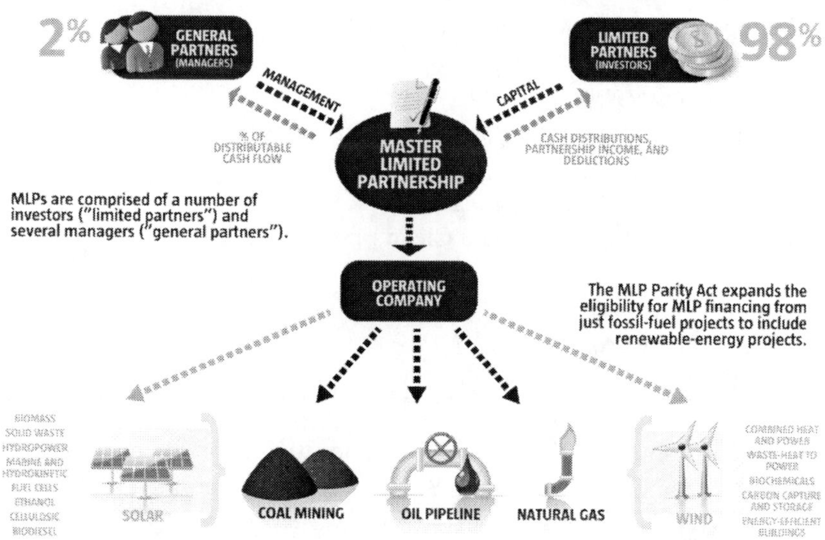

The MLP Parity Act has been endorsed by

Businesses

Duke Energy
DuPont
Imperium Renewables FAIR Coalition
FloDesign Wind Turbine Leucadia Energy, LLC Myriant Corporation NRG
OwnEnergy
Renewable Biofuels, Inc. Summit Power Group Virent

Trade Associations

Advanced Biofuels Association Advanced Ethanol Coalition
Alliance for Industrial Efficiency
American Biogas Council Biomass Power Association Energy Recovery Council Energy Storage Association Geothermal Energy Association Growth Energy Heat is Power Association International District Energy Association National Enhanced Oil Recovery Initiative
Ocean Renewable Energy Association Offshore Wind Development Coalition Real Estate Roundtable Semiconductor Equipment and Materials International Solar Energy Industries Association

Environmental Advocates

Clean Air Task Force Environmental and Energy Study Institute

Think Tanks and Research Institutes

Clean Energy Trust New England Clean Energy Council
Third Way

Finance & Investment

CalCEF, CERES

Writing in the New York Times in June 2012, Dan Reicher and Felix Mormann of Stanford University's Steyer-Taylor Center for Energy Policy and Finance described the appeal of MLPs: "Master limited partnerships carry the fund-raising advantages of a corporation: ownership interests are publicly traded and offer investors the liquidity, limited liability and dividends of classic corporations. Their market capitalization exceeds $350 billion. With average dividends of just 6 percent, these investment vehicles could substantially reduce the cost of financing renewables."

Because MLPs are so attractive to investors, they have been proven to bring new capital into American energy projects. This is especially important in the case of renewable-energy generation, where it is harder for investors to see as quick a return as compared to fossil fuel-based energy generation, for which much of the processing and transportation infrastructure was built decades ago. Constructing the same level of critical infrastructure for renewable energy sources will take time and investment, so the MLP Parity Act levels the playing field and helps address that problem.

An MLP must generate at least 90 percent of its income from qualified sources, such as real estate or natural resources, including crude oil, natural gas, petroleum products, coal, timber, and other minerals. Section 613 of the federal tax code specifically requires qualifying energy sources to be depletable resources – meaning we are working against our own goal of an "all of the above" energy strategy that includes additional homegrown renewable energy sources.

What are the benefits of an MLP?

Pass-through tax structure (no double taxation) leaves more cash to distribute to investors

Cash distributions are mostly tax-deferrable

Lower cost of capital provides MLPs with advantages in building and acquiring assets

Public trading allows MLPs to raise capital from a broader range of investors than if not traded

Can own rate-regulated assets and still give investors an attractive rate of return

	MLP	LLC	Corporation
Taxable at entry level	No	No	Yes
Tax items flow through	Yes	Yes	No
Tax deferral on distributions	Yes	Yes	No
Tax reporting	K-1	K-1	DIV-1099
General partner	Yes	No	No
IDRs	Yes	No	No
Investor voting rights	No	Yes	Yes

Source: National Association of Publicly Traded Partnerships

How the MLP Parity Act Helps

The *MLP Parity Act* simply expands the definition of "qualified" sources to include clean energy resources and infrastructure projects. Specifically included are those energy technologies that qualify under Sections 45 and 48 of the tax code, including wind, closed and open loop biomass, geothermal, solar, municipal solid waste, hydropower, marine and hydrokinetic, fuel cells, and combined heat and power.

The legislation also allows for a range of transportation fuels to qualify, including cellulosic, ethanol, biodiesel, and algae-based fuels, as well as energy-efficient buildings, electricity storage, carbon capture and storage, renewable chemicals, and waste-heat-to-power technologies.

The *MLP Parity Act* does not affect any current MLP entity. All projects currently eligible to structure as MLPs would continue to qualify exactly as they would under existing law.

History of Master Limited Partnerships

The first MLP was created in 1981 to attract capital by offering small investors a partnership investment in a liquid security. The success of the Apache Oil Company MLP led to other oil and gas MLPs, as well as MLPs formed for other capital-dependent enterprises.

Congress first established rules for master limited partnerships in 1987 legislation that introduced Internal Revenue Code Section 7704 and, for the first time, defined "publicly traded partnerships." The MLP structure was limited to businesses deriving 90 percent of their income from specific sources, including dividends, rents, interests, capital gains, and mining and natural resources income identified in Section 613 of the tax code. This definition allowed oil and gas extraction and transportation activities access to the MLP structure, but excluded other energy sources.

In 2008, the *Emergency Economic Stabilization Act* (P.L. 110-343) expanded the definition of qualifying income to include transportation and storage of certain renewable and alternative fuels (ethanol, biodiesel, and a series of liquefied fuels), as well as industrial-source carbon dioxide.

The National Association of Publicly Traded Partnerships estimates there are more than 100 MLPs currently being traded on major exchanges, primarily focused on energy-related industries and natural resources. "Midstream" oil and gas projects — gathering, processing, pipelines, and distribution — account for the majority of current MLPs.

Of the estimated $445 billion in MLP capital currently in the market, approximately $400 billion (89 percent) has gone into qualifying energy and natural resources. Of that, just under 80 percent has gone into midstream oil and gas pipeline projects.

BROAD SUPPORT FOR THE MLP PARITY ACT

David Crane, President, NRG Energy: 'The MLP Parity Act is a phenomenal idea. It's a fairly arcane part of the tax law, but it's worked well and has been extremely beneficial to private investment in the oil and gas space. The fact that it doesn't currently apply to renewables is just a silly inequity in our current law."

Josh Freed, Vice President for Clean Energy, Third Way: 'The MLP Parity Act helps take government out of the role of picking energy winners and losers. Master Limited Partnerships have a long history of success raising private capital to finance oil and gas pipelines in the United States. This relatively small change to the tax code empowers investors, rather than Congress, to determine what kind of energy projects make financial sense. '

Jim Lanard, President, Offshore Wind Development Coalition: 'MLPs will help to support the establishment of a sustainable offshore wind industry, since they can reduce the cost of capital and attract more investors. While extension of the Investment Tax Credit (ITC) remains our industry's number one legislative priority, MLPs will serve as a nice complement to ITCs."

Brooke Coleman, Executive Director, Advanced Ethanol Council: 'The advanced ethanol industry strongly supports the efforts of Senator Coons and the co-sponsors of Master Limited Partnership Parity Act to level the playing field for advanced technologies when it comes to MLPs. Oil and gas producers are using MLPs to access the retail investment market, which in turn makes it easier to finance new oil and gas projects. Advanced, renewable technology developers are explicitly excluded from eligibility for MLPs. It makes no sense for the federal government to continue to offer this financing vehicle to fossil fuels only. The MLP Parity Act would take a meaningful inequity out of the federal tax code and put the country in a better position to create jobs and compete in the emerging $2 trillion global clean energy marketplace."

Robert P. Thornton, President & CEO, International District Energy Association: 'Since district energy and combined heat and power systems are capital intensive investments, the Master Limited Partnership Parity Act offers significant potential to advance the growth of these highly efficient and resilient clean energy resources. By providing a new source of liquid private capital, the Act can play a vital role in strengthening the infrastructure of our nation's cities, communities, institutions and military bases."

Tony Straquadine, Chairman, The Heat is Power Association; Manager of Government Affairs, Alliance Pipeline: 'The waste heat to power industry strongly supports the efforts of Senators Coons and Moran and the other co-sponsors of the Master Limited Partnership Parity Act to ensure that projects using a broad array of energy generation technologies can avail themselves of the MLP structure. The ability of waste heat to power projects to qualify for MLPs will make those projects easier to finance, will be attractive to a broader range of energy investors, and will produce emission-free power from an otherwise wasted resource. We applaud your efforts to level the playing field for energy generation resources like ours that improve the competitiveness of our nation's industrial sector and generate power with no combustion and no emissions."

Collin O'Mara, Secretary, Department of Natural Resources and Environmental Control, State of Delaware: 'Senator Coon's vision to expand Master Limited Partnerships to include projects that harness all energy resources simply makes sense. At a time when we need to spur private sector investment, create domestic jobs, transition to a clean energy future, and fulfill our climate responsibilities, what better tool can we unleash than expanding upon a tried and true financing vehicle that has spurred billions of dollars of energy investments over the past decades and helped bring energy resources to all corners of our nation. By spurring private investment into large-scale renewable energy projects, like solar, fuel cells, geothermal, and offshore wind, we will drive down the costs of projects and catalyze greater innovation, commercialization, and deployment of the very technologies that are critical to our nation's economic and environmental future."

Jim Collins, President, DuPont Industrial Biosciences: 'We appreciate the leadership of Senator Coons and his Senate cosponsors and Representative Poe and his House cosponsors on the Master Limited Partnership Parity Act. This constructive bicameral, bipartisan legislation will facilitate the commercialization of advanced renewable fuel technologies by extending the tax-efficient Master Limited Partnership structure to these investments. These tax policies have proven effective in encouraging investment in oil and gas infrastructure and can similarly help in the domestic advanced biofuels industry, bringing investment and jobs to rural America while expanding domestic energy production and reducing our vulnerability to oil price shocks."

Ted Michaels, President, Energy Recovery Council: 'The Energy Recovery Council commends Senator Coons, Senator Moran, and the sponsors for introducing this important legislation which will provide renewable technologies with access to much needed capital. This legislation is another critical step in recognizing the value of renewable energy and ensuring that technologies such as waste-to-energy can continue to pursue growth opportunities."

Kurt Waltzer, Special Projects Director, Clean Air Task Force: 'The Clean Air Task Force strongly supports the MLP Parity Act. Carbon Capture and Storage is a critical path technology for addressing climate change and we need more pioneer projects. If adopted, this bill will cut the cost of getting that steel in the ground".

Judi Greenwald and Brad Crabtree, Co-Directors, National Enhanced Oil Recovery Initiative: 'On behalf of participating industrial and other companies, labor unions, environmental organizations and state officials, the National Enhanced Oil Recovery Initiative applauds the MLP Act's sponsors for expanding MLP eligibility to increase private investment in carbon capture and storage projects. Capturing carbon dioxide from power plants and many other industrial facilities for use in enhanced oil recovery produces more American oil, creates good-paying jobs, generates net new revenue for the federal treasury, and reduces CO2 emissions."

Rhone Resch, president and CEO, Solar Energy Industries Association: 'This bill is an important step toward leveling the playing field between clean, renewable energy and long-entrenched energy sources in America, by providing the solar industry with access to private capital in the same manner enjoyed by the oil and gas industry for almost 30 years. Today, the U.S. solar industry employs nearly 120,000 Americans, with solar deployment growing and costs to consumers dropping. Senator Coons' MLP proposal would build on this success, and SEIA applauds him for putting forward an idea that has the potential to attract private sector investment in critically important solar projects. This bill represents smart public policy, and solar represents a clean, safe, affordable and inexhaustible supply of energy for our nation. We look forward to working with Senator Coons and other stakeholders on this important issue."

Vice Admiral Dennis McGinn (USN-Ret.), President and CEO, American Council On Renewable Energy: 'We commend Senator Coons for his leadership in introducing important legislation to level the playing field and promote greater private investment in our nation's abundant and affordable renewable energy resources and fuels. Enabling master limited partnership investment in renewable energy and infrastructure can help lower project costs, leading to more economic investment and a more diverse energy mix."

David Gardiner, Executive Director, Alliance for Industrial Efficiency: 'We applaud Senators Coons (D-DE), Moran (R-KS), Stabenow (D-MI) and Murkowski (R-AK) for their bi-partisan leadership in introducing the Master Limited Partnership Parity Act. The Alliance for Industrial Efficiency is particularly pleased that the bill extends low-cost financing to Combined Heat and Power and Waste Heat to Power, proven clean energy sources that could provide as much as 20 percent of U.S. electric capacity."

Mike McAdams, President, Advanced Biofuels Association: 'We are grateful for Senator Coons' leadership at a critical point for America's domestic biofuels industry as we are moving from the beaker to the barrel, in record time. The legislation provides an innovative financial mechanism that could significantly reduce the cost of financing as companies are reaching a game-changing milestone. Substantial investments by private companies in research and development have been the catalyst for today's success in bringing advanced biofuels to commercial markets, but stable and consistent public policies are crucial to encourage and allow additional investment dollars that will help get us across the finish line. By creating a new and more appealing option for investors, the bill helps level the playing field and ultimately promotes a more cost competitive advanced biofuel alternative to conventional fuel."

Rob Gramlich, Interim CEO, American Wind Energy Association: 'America's wind energy sector is a success story that has proven its strength by recruiting $18 billion in annual private investment in America's energy infrastructure in recent years despite short-lived policy certainty. We commend Senators Chris Coons, Jerry Moran, Debbie Stabenow, Lisa Murkowski and Representatives Ted Poe, Mike Thompson, Peter Welch and Chris Gibson for their leadership in promoting the eligibility of master limited partnerships to include renewable energy projects. MLPs work well for conventional energy infrastructure and will be one critical tool to ensure that federal incentives continue to drive private investment into the U.S. wind energy market. We look forward to working with Senator Coons, Representative Poe, and the other bill co-sponsors to proceed with legislation that allows wind power developers to efficiently utilize MLP structures."

In: Master Limited Partnerships …
Editor: Josiah Heslin

ISBN: 978-1-63117-012-6

Chapter 4

MASTER LIMITED PARTNERSHIPS: A POLICY OPTION FOR THE RENEWABLE ENERGY INDUSTRY*

Molly F. Sherlock and Mark P. Keightley

SUMMARY

Expanded investment in clean and renewable energy resources continues to be a policy priority of the Obama Administration and an area of interest to the Congress. In recent years, the primary policy vehicle for promoting investment in renewable energy has been tax credits, particularly the renewable energy investment and production tax credits. A lack of tax liability, however, has limited the renewable energy sector's ability to fully take advantage of these and other tax benefits. The result has been an increased interest in exploring other options for promoting investment in renewable energy. One option might be to allow renewable energy entities access to the master limited partnership (MLP) business organizational form.

An MLP is a type of business structure that is taxed as a partnership, but whose ownership interests are traded on financial markets like corporate stock. Being treated as a partnership for tax purposes implies that MLP income is generally subject to only one layer of taxation in contrast to publically traded C corporations, which are subject to two

* This is an edited, reformatted and augmented version of a Congressional Research Service publication, CRS Report for Congress R41893, prepared for Members and Committees of Congress, from www.crs.gov, dated June 28, 2011.

layers of taxation. The ability to access equity markets in a manner similar to corporations allows MLP to obtain greater amounts of capital. Access to a greater pool of capital, when combined with the favorable partnership tax treatment, may allow MLPs to secure capital at a lower cost than similar businesses operating under a different organizational structure. The lower cost of capital, in turn, could increase investment in the renewable energy sector. Congress first established rules relating to MLPs in the 1980s. At that time, the MLP structure was limited to businesses deriving 90% of their income from primary sources, which included dividends, interest, rents, capital gains, and mining and natural resources income. Effectively, this definition allowed oil and gas extraction and transportation activities access to the MLP structure, while renewable energy resources were generally excluded. The Emergency Economic Stabilization Act of 2008 (P.L. 110-343) expanded the definition of income from qualifying sources to include transportation of certain renewable and alternative fuels, such as ethanol and biodiesel.

Provisions enacted under the American Recovery and Reinvestment Act of 2009 (ARRA; P.L. 111-5) enhanced the value of certain renewable energy tax credits for many renewable energy projects. By transforming existing tax credits into cash grants, the Section 1603 grants in lieu of tax credits program enhanced access to capital for many in the renewable energy sector. The Section 1603 grant program is scheduled to expire at the end of 2011. As Congress evaluates other policies to attract additional capital to the renewable energy sector, allowing renewable energy entities to structure as MLPs might be one option.

Extending the MLP structure to renewables could possibly attract additional capital to and stimulate investment in the renewable energy sector. There are, however, a number of potential policy concerns to consider. First, expanding access to the MLP structure could narrow the corporate tax base, which is one of the reasons access to this structure was limited in the first place. Second, if changes to the tax code allowing renewable entities to access the MLP structure are enacted alongside changes to current passive activity loss rules, there may be concerns about the possibility of renewable energy investments being used as a tax shelter. Finally, if the concern is that renewable entities are disadvantaged relative to fossil fuels currently able to use the MLP structure, one option would be to prevent other energy entities from structuring as MLPs.

INTRODUCTION

Ongoing concern surrounding energy security and the environment have led to sustained Congressional interest in energy tax policy. The 111th Congress enacted a number of renewable energy tax incentives as part of the

American Recovery and Reinvestment Act of 2009 (ARRA; P.L. 111-5). A number of expiring renewable energy tax provisions were extended through the end of 2011 in the Tax Relief, Unemployment Insurance Reauthorization, and Job Creation Act of 2010 (P.L. 111-312). As various incentives for renewable energy, renewable fuels, and energy efficiency are scheduled to expire at the end of 2011, the Congress may want to explore various policy options for encouraging investment in renewable energy.

The Obama Administration has also repeatedly emphasized the importance of investments in clean energy technologies and infrastructure. President Obama has noted that clean energy investments can enhance domestic energy security, promote environmental objectives, and create jobs.[1] One barrier to investments in renewable energy projects is that such projects are highly capital intensive. Capital intensive renewable energy projects continue to face a number of challenges with respect to financing. Policymakers have been exploring various options for increasing the availability and decreasing the cost of financing for the renewable energy sector. One option for attracting additional capital to the renewable energy sector that Congress may consider is allowing renewable energy activities access to the master limited partnership (MLP) business structure.[2]

This report explores the policy option of extending the master limited partnership (MLP) business structure option to renewable energy facilities and related activities. Before evaluating the policy, this report provides a brief overview of the MLP structure, highlighting notable tax issues. This report also provides background on the legislative origins of the MLP structure, and legislative changes affecting MLPs that have been made in recent years. The final sections of this report highlight how the MLP structure might be able to attract additional capital to the renewable energy sector, while also discussing some potential policy concerns.

MASTER LIMITED PARTNERSHIPS

What Is an MLP?

A master limited partnership (MLPs) is a type of business structure that is taxed as a partnership, but whose ownership interests are traded on financial markets like corporate stock.[3] Being treated as a partnership for tax purposes implies that MLP income is generally subject to only one layer of taxation. Income passes through the partnership to its business owners who pay taxes according to the individual income tax system.[4] Publically traded C corporations, however, are subject to two layers of taxation. Their earnings are taxed once at the corporate level, according to the corporate tax system, and then a second time at the individual-shareholder level when dividend payments are made or capital gains are realized.[5] This leads to the so-called "double taxation" of corporate profits.

Businesses may be able to attract more capital at a lower cost by choosing to organize as an MLP than would otherwise be possible. Ownership interests of MLPs, which are known as units to distinguish them from corporate stock, are traded on regulated financial exchanges in the same manner as shares of corporate stock. MLP units are an attractive alternative to corporate stock for individual investors, since the single layer of taxation can result in higher after-tax returns. The higher returns to investors corresponds to lower financing costs for the business. And the ability to have their units traded on public exchanges is attractive to the MLP itself since it grants them access to larger and more liquid sources of capital, which can be used to pursue investments.

MLPs are typically formed as a limited partnership consisting of thousands of limited partners and at least one general partner.[6] The limited partners are public investors who provide most of the capital to the MLP in exchange for publically tradable units. MLP units pay a periodic cash distribution similar to a dividend and are traded on the New York Stock Exchange, NASDAQ, American Stock Exchange, or over-the-counter market, like stocks. Unit holders are also allocated a portion (determined by units owned) of the partnership's income, deductions, and credits. When a unit is traded, the buying investor replaces the selling investor as a limited partner in the MLP. Currently, there are over 100 publically traded MLPs, the majority of which are energy related (see *Appendix).*

The MLP's general partner manages the partnership in exchange for a percentage of the partnership's income, called an incentive distribution right (IDR). The exact percentage that the general partner will receive is agreed to when the MLP is formed, but typically involves a 2% share of some baseline amount of distributable cash flow. The general partner may then receive an increasing share of distributable cash flow above the baseline if the yield on the limited partner's units exceed certain thresholds. This payment structure is thought to compensate the general partner for taking on certain risks and encourage them to manage the MLP in a way that maximizes the return to investors. The general partner may be another (parent) company or a group of individuals.

To be an MLP at least 90% of a business's gross income must be considered "qualifying income."[7] Qualifying income generally includes dividends, interest, rents, capital gains, and mining and natural resource income. [8] Income related to the exploration, development, mining or production, processing, refining, transportation, storage, and marketing of any mineral or natural resource falls under the latter income category.9 Recently, the definition of qualifying income was expanded. The expanded definition includes income from the transportation and storage of certain renewable and alternative fuels, including cthanol and biodiesel, and activities involving industrial source carbon dioxide.[10]

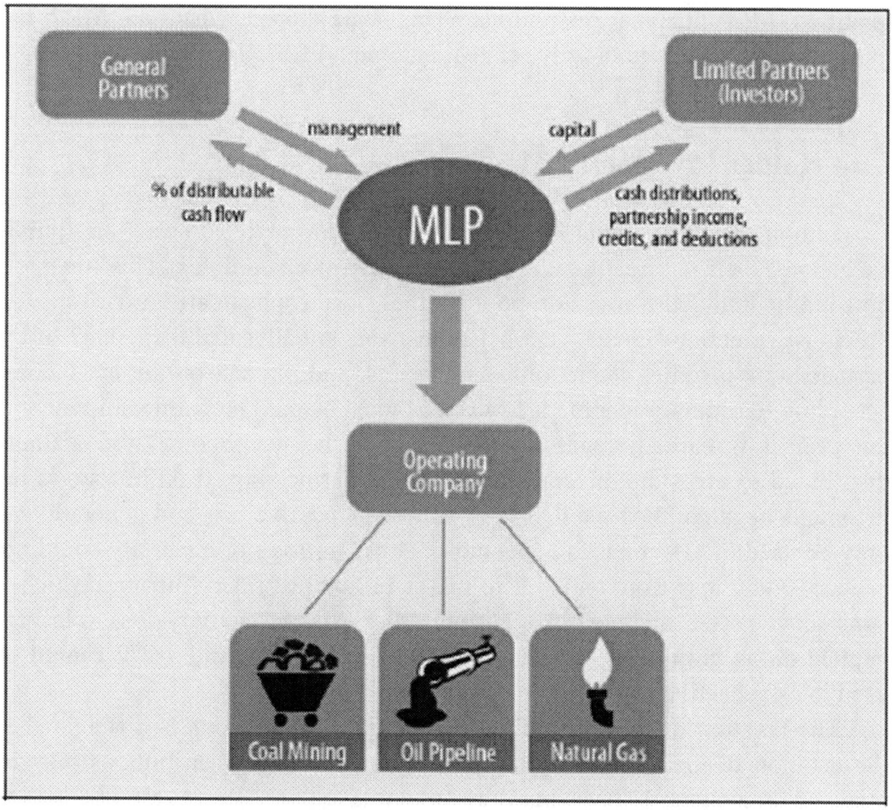

Source: CRS.

Figure 1. Basic MLP Structure.

MLPs will typically own and operate their actual business assets indirectly through a subsidiary, known as an operating company. Historically, owning business assets indirectly through a separate company reduced the administrative burden associated with MLPs being publically traded.[11] At one point, direct ownership of the assets might have required that an MLP file change of ownership documents in every state it had operations in whenever investors traded shares.

Today, operating companies are often used by MLPs to limit liability among various ventures operating in different states. The use of an operating company also gives MLPs more options and protection when structuring its debt financing since it can subordinate and separate debt along its various lines of business more easily. Additionally, MLPs can use an operating company to "filter" income generated by a subsidiary that would otherwise violate the qualified MLP income restrictions. This can happen when an MLP has business investments in closely related, but unqualified, lines of business.

Unit Holder Investor Tax Issues

Although MLP units trade alongside corporate stocks on public exchanges, their tax treatment is fundamentally more complex. This treatment potentially limits the investor pool to the most sophisticated investors. As previously mentioned, each year MLP investors are allocated their share of the partnership's income, deductions, and credits, and pay tax on the net income according to ordinary income tax rate. Tax must be paid on partnership income the year it is earned, regardless of whether the net income was actually distributed to investors or retained within the partnership. If an investor's net income is negative then the loss is considered a passive loss and generally can only be used to offset passive income.[12] Thus, an investor generally could not use a $10,000 loss from an MLP to offset $10,000 of salary income, which is considered active income. In addition, MLP passive activity loss rules are applied on an entity-by-entity basis.[13] Thus, losses from one MLP cannot be used to offset active income from another MLP.

The taxation of an investor's periodic cash distributions is different than the taxation of their share of partnership income. Cash distributions may be received on a quarterly basis but they are not taxed until the investor sells their MLP units. Furthermore, when the distributions are taxed, they are generally taxed as capital gains and not ordinary income. To compute the capital gain related to distributions an investor will begin with the unit sales price and subtract their adjusted basis in the MLP. An investor's adjusted basis is the original purchase price of the units decreased by the amount of cash distributions received and increased by their share of the partnership's net income. The basis adjustments ensure that all income is only subject to one layer of taxation. Investors may also have to make various other tax computations when they sell units, including determining their share of depreciation deductions that are subject to recapture (taxation).[14]

Another potential tax complexity that may limit who invests in MLPs is unrelated business income taxation (UBIT). The term "unrelated business income" generally means income generated by a tax-exempt organization that is unrelated to the organization's regular business.[15] With regard to MLPs, the issue of UBIT primarily concerns pension funds and tax-preferred accounts such as IRAs and college savings plans that invest in MLPs. These tax-preferred investment and saving vehicles must recognize any income greater than $1,000 earned from MLP investments as unrelated business income and pay tax on that income (UBIT). Unrelated business income is taxed according to the corporate rate schedule. The effect on the after-tax return for pension funds and individuals using IRAs may limit these investors in MLPs.

MLPs and Carried Interest

The President's 2010, 2011, and 2012 Budget Outlines along with numerous other proposals in prior Congresses proposed changes to the tax treatment of certain types of "carried interest," which some have been concerned would impact energy-related MLPs. Carried interest refers to the percentage of a partnership's earnings that its general partners (managers) receive as a performance fee. Proposals regarding carried interest have generally been aimed at the financial services industries. Hedge funds and private equity funds, which are typically set up as a partnership, pay management a fee that depends on the performance of the fund.[16] This fee, which represents carried interest, is taxed at the more favorable capital gains rate instead of the ordinary income rates since these firms are involved in the buying and selling of financial assets, which results in capital gains. Past Administration and congressional proposals would have taxed carried interest as ordinary income, stating that the payments represent compensation for services and not capital.

Attempts to change the tax treatment of carried interest, currently treated as capital gains, would likely have minimal impact on current and future energy-related MLPs. It is true that the IDR payments made to MLP general partners are a form of carried interest since they represent a fee based on the performance of the partnership. These payments, however, are the result of operating business income and not capital gains income from the buying and selling of assets, as is the case with a hedge fund. As a result, IDR payments are already taxed mostly at ordinary income rates. There may be a small fraction of MLP carried interest related to the occasional sale of capital assets used in the operating business that is taxed as a capital gains right, but this amount is likely minimal.

LEGISLATIVE HISTORY

Master limited partnerships first appeared in the early 1980s.[17] The Tax Reform Act of 1986 (TRA86) reduced the top marginal individual income tax rate to a level lower than the top marginal corporate tax rate. As a result, the partnership business structure became more favorable for tax purposes than the corporate structure. Other changes enacted as part of the TRA86, however, limited the attractiveness of MLPs for investors. Specifically, the TRA86 introduced passive loss rules, which prevented investors from using deductions associated with businesses in which they are not actively involved, such as MLPs, to offset other types of income.

In 1987, Congress enacted IRC § 7704, which modified the rules publicly traded partnerships (PTPs) and MLPs had been using to avoid being subject to corporate taxation.[18,19] Under the new rules, partnerships whose ownership interests were publically traded were to be treated as corporations for tax purposes. An exception was made, however, that allowed partnerships meeting two criteria to continue being taxed as partnerships. The two criteria were (1) the partnership was in existence on December 17, 1987, and (2) at least 90% of its gross income came from passive sources, such as rents, royalties, and natural resource income, among others. If the income criteria was not met, the partnership could be grandfathered for a 10-year period, after which it would be taxed as a corporation or, in the extreme, cease operations.

The 1987 rules related to PTPs and MLPs were enacted to address concerns surrounding erosion of the corporate tax base. In the House Report accompanying H.R. 3545, the 100th Congress noted "To the extent activities that would otherwise be conducted in the corporate form, and earnings that would be subject to two levels of tax (at the corporate and shareholder levels), the growth of publically traded partnerships engaged in such activities tends to jeopardize the corporate tax base."[20] Further, Congress also observed that PTPs had been used to avoid corporate taxes, noting that the intent of pre-1987 tax law was "being circumvented by the growth in publically traded partnerships that are taking advantage of an unintended opportunity for disincorporation and elective integration of the corporate and shareholder levels of tax."[21]

When IRC § 7704 was enacted, effectively subjecting most PTPs and MLPs to corporate taxation, existing PTPs and MLPs were allowed to continue operating as a partnership for 10 years. In 1997, legislation was passed that allowed PTPs and MLPs that had been grandfathered and allowed to continue operating as partnerships an additional choice.[22] Instead of being forced to choose an alternative organizational form, grandfathered PTPs and MLPs were given the option of paying a 3.5% tax on gross income, as an alternative to corporate income taxes.

Legislative changes enacted as part of the American Jobs Creation Act of 2004 (P.L. 108-357) potentially expanded the pool of capital able to invest in MLPs. Provisions in this legislation effectively changed rules related to UBIT, which had previously made it unattractive for mutual funds to invest in MLPs. Specifically, the 2004 Jobs Act allowed partnership distributions to be considered qualifying income for mutual funds, thus allowing funds to invest in MLPs without having to worry about UBIT. This change effectively increased the potential pool of MLP investors.

Most recently, the definition of qualifying MLP income was expanded to include the transportation and storage of certain renewable and alternative fuels, including ethanol and biodiesel, and other activities involving industrial source carbon dioxide.[23] This change was made as part of the Emergency Economic Stabilization Act of 2008 (P.L. 110-343), at an estimated cost of $119 million over 10 years.[24] The purpose of this change was to allow biofuels pipelines to receive the same tax treatment as petroleum pipelines.[25] Previously, the statutory definition of qualifying income had not been expanded since § 7704 was added to the Internal Revenue Code in 1987.

MLPs and Renewable Energy

As noted above, policies enacted in the 111th Congress were designed to support investment and growth in the renewable energy sector. These policies were consistent with the objectives of the Obama Administration, which has emphasized the importance of investments in clean energy technology, including resources and infrastructure.[26] Enhanced use of clean energy resources may be consistent with broader energy policy goals, environmental sustainability, and perhaps domestic energy security.[27] Enhanced investment in and deployment of renewable energy technologies may also have the potential for domestic job creation.[28]

During the 111th Congress, action was taken to address certain challenges in financing renewable energy projects. In the wake of the recent financial crisis, the renewable energy sector was faced with new challenges in financing investment. Prior to 2008, renewable energy investors often relied on tax-equity markets to monetize renewable energy tax benefits (such as the renewable energy production tax credit (PTC)).[29] The ability to monetize federal renewable energy tax incentives was important for renewable energy investors to finance these capital intensive projects. Tax-equity financing became increasingly scarce during 2008 and 2009, leading Congress to enact the Section 1603 grants in lieu of tax credits program.[30] Under this program, qualified taxpayers could elect to receive a one-time grant from the Treasury in lieu of the renewable energy PTC or investment tax credit (ITC). The grant option effectively eliminated the need for tax-equity partnerships for many eligible taxpayers.

It has been argued that the Section 1603 grant program prevented what could have been a substantial decline in renewable energy investments, and may have resulted in additional investments in renewable energy generation capacity.[31] Allowing taxpayers to receive a direct grant from the Treasury, and avoid the tax-equity market, has been credited with broadening the pool of renewable energy investors.[32] The success of the Section 1603 program has led some to note the potential value of policies that will attract additional capital to the renewable energy sector. One policy option that proponents note might attract capital to the renewable energy sector would be to allow renewable energy developers to structure as a master limited partnership(MLP).[33,34]

Should Congress decide to expand the definition of qualifying income to include renewable energy, or make other changes to current tax laws that would allow renewable energy entities to structure as MLPs, Congress may decide to stipulate which clean or renewable energy activities would qualify. In the case of the Section 1603 grant program, qualifying renewable energy technologies were those that were already eligible for the renewable energy PTC or ITC. The MLP structure could be extended to renewable energy technologies already eligible for other renewable energy tax incentives, or expanded to include other technologies that might support expanded use of renewable energy, such as advanced energy storage and transmission technologies.

Extending the MLP Structure to Renewables: Potential Benefits

Proponents of extending the MLP structure to renewable electricity generation facilities note that doing so might help attract additional capital to the sector. Additional capital would be attracted to the sector by the higher returns typically offered to investors by MLPs. Additionally, allowing renewable power facilities to structure as MLPs could provide easier access to equity. Being able to sell shares to raise equity is a benefit typically reserved for C-corporations. Given these advantages, MLPs might allow clean energy projects to produce energy at a lower cost, and thus be more competitive with fossil fuels, including coal and natural gas.

The MLP structure might also help attract investors to the renewable energy sector, particularly if changes allowing renewables to structure as MLPs were enacted alongside changes to existing passive loss rules. Without accompanying changes to the passive loss rules, the benefits associated with the MLP structure are more limited. As noted above, renewable energy taxpayers have historically turned to the tax-equity market to monetize renewable energy tax incentives (especially prior to the enactment of the Section 1603 grant program in 2009). Under the MLP structure, tax losses pass through to investors. If passive loss rules are restructured to allow investors to use these tax losses to offset other income, renewable energy investments might become more attractive.[35] From this perspective, the renewable energy entity might be able to attract additional capital to the sector with the MLP structure since such a structure could be designed to allow investors to directly benefit from renewable energy tax incentives. While modifications to the passive loss rules would help maximize the benefits investors are able to realize from the MLP structure, extending the MLP structure to renewables without changing the passive loss rules may also provide some benefit, as MLPs are not subject to corporate level taxation and can raise capital by selling additional shares.

Allowing renewables to structure as MLPs may help reduce the cost of capital, thereby increasing investment in renewable energy. While expanding the pool of investors and increasing access to capital may help promote investment in renewable energy, the overall value of the ability to structure as an MLP is likely small relative to existing tax benefits for renewable energy entities. Thus, this has led some to observe that "Because MLP s would only increase the eligible investor pool ... by themselves they would most likely not supplant the tax incentives currently in place."[36]

Extending the MLP Structure to Renewables: Potential Policy Concerns

Expanding the definition of qualifying income to allow renewable energy producers to structure as MLPs could raise concerns with respect to the size of the corporate tax base. As was noted above, the rules generally treating MLPs as corporations were enacted in 1987 to address concerns about the long-term erosion of the corporate tax base.[37] Expanding the definition of qualifying income and allowing more firms to structure as MLPs would likely result in a decrease in the size of the corporate tax base, and result in federal revenue losses.[38] Under current law, the tax expenditure (revenue loss) associated with allowing certain energy companies to structure as MLPs is estimated to be around $2.8 billion over the 2010 through 2014 time period.[39]

The increasing proportion of income flowing through passthroughs as opposed to corporate entities has caught the attention of the Administration and Congress. Senator Max Baucus, chairman of the Senate Finance Committee, in the context of tax reform, has noted "We're going to have to look at passthroughs—say they've got to be treated as corporations if they earn above a certain income. It's one possibility."[40] Treasury Secretary Timothy Geithner has also made reference to the issue of allowing corporations alternative organizational forms, stating "I think, fundamentally, Congress has to revisit this basic question about whether it makes sense for us as a country to allow certain businesses to choose whether they're treated as corporations for tax purposes or not."[41] While much of this concern has been directed at large passthrough entities (those with $50 million or more in revenues), it may still be important to consider when it is appropriate to allow entities to structure as passthroughs, alongside the associated revenue cost.

If the concern is that allowing fossil fuel energy entities to structure as MLPs puts renewables at a disadvantage, preventing publicly traded fossil fuel entities from structuring as passthrough entities is another option. Requiring publicly traded companies (energy and other types of MLPs) to structure as C-corporations would broaden the corporate tax base, and ensure that such fossil-fuels energy-related MLPs do not have better access to capital or a preferred tax status not available to renewable energy alternatives. This policy change, however, could place greater capital constraints on, and potentially reduce investment in, industries currently able to use the MLP structure.

Allowing renewable energy facilities to structure as MLPs, if enacted jointly with policies that would exempt renewable energy tax benefits from passive activity loss rules, could raise concerns surrounding "gold plating" of renewable energy projects or the possible use of tax shelters. Gold plating can occur when investors look to invest in renewable energy property for the purpose of tax benefits without regard to performance and production. Specifically, if investors are able to use renewable energy tax benefits to offset active income from other sources, the potential for tax shelter opportunities may emerge. During the 1980s, investors using debt to finance large wind projects could generate tax benefits through investment tax credits that were available at the time. The tax benefits were valuable in and of themselves, even if the wind facility did not produce electricity.[42,43] Removing passive activity loss rules for renewables eligible for generous investment tax credits could create opportunities for tax shelters like those seen in the 1980s.

One final point of potential concern is that MLPs have typically been used to finance proven technologies with stable cash flows. Since the financing structure is particularly well suited to entities with predictable cash flows, many existing MLP operations are involved in transportation of fuels or other midstream operations. Renewable energy technologies that pose technology risk may not be well suited to take advantage of the MLP structure. Capital is most scarce for energy technologies that have been developed beyond the research & development (R&D) laboratory phase, but have not yet reached commercialization.[44] MLPs are not likely to attract additional capital to this capital-scarce sector comprised of technologies that have moved beyond field testing but have not yet been deployed at scale.

CONCLUSION

Additional access to capital has the potential to stimulate investment and growth in the renewable energy sector. MLPs could have the potential to attract additional capital to the renewable energy sector. MLPs could also allow investors to benefit from other renewable energy tax incentives, and thereby avoid tax-equity markets for monetization of renewable energy tax benefits, if changes to existing passive loss restrictions were enacted.

Extending the definition of qualifying income to allow renewable energy facilities to structure as MLPs might raise policy concerns. Specifically, expanding the definition of MLPs to include other types of activity could be viewed as a narrowing of the corporate tax base. Further, if changes in current law regarding qualifying income under the MLP structure are coupled with changes in passive activity loss rules, there are concerns that such changes could lead to tax shelter opportunities.

Appendix. MLPs by Industry: Past and Present

The MLP universe has grown and changed in recent decades. In the energy sector alone, the number of energy MLPs increased from 6 in 1994 to 72 in 2010.[45]

Over that same time period, total market capitalization of energy MLPs grew from $2 billion to roughly $220 billion.[46] Additionally, since the 1990s, the universe of MLPs has changed to include a larger proportion of energy MLPs, specifically those involved in midstream operations.

Figure A-1 illustrates the proportion of MLPs in different industry groups in 1990 and 2010. Over the 20-year period, the share of MLPs in the energy industry has increased. Further, energy MLPs have become more likely to be involved in midstream or transportation activities over time, as opposed to extraction and production.

In 1990, 10% of MLPs were oil and gas midstream operations. By 2010, this share had increased to 44%. Over the same period, the proportion of MLPs involved in oil and gas exploration and production decreased from 21% to 10%.

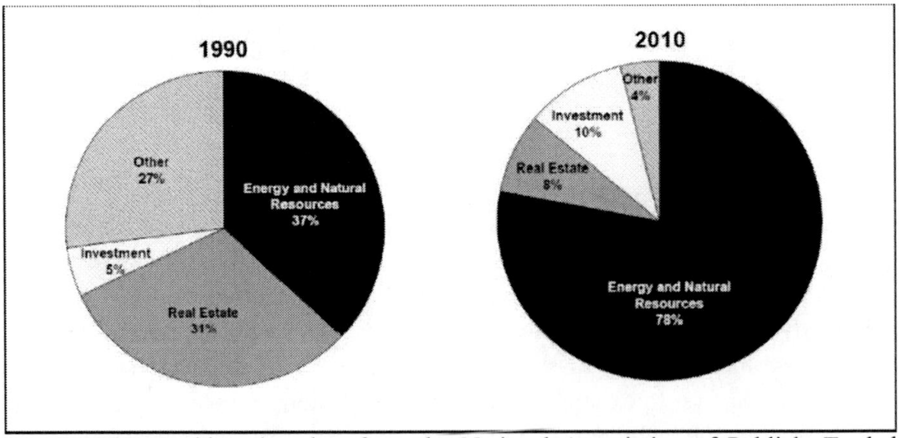

Source: CRS graphic using data from the National Association of Publicly Traded Partnerships, available at http://www.naptp.org/documentlinks/Investor_Relations/MLP_101.pdf.

Notes: Real estate includes income properties, homebuilders, and mortgage securities.

Figure A-1. MLPs by Industry Group.

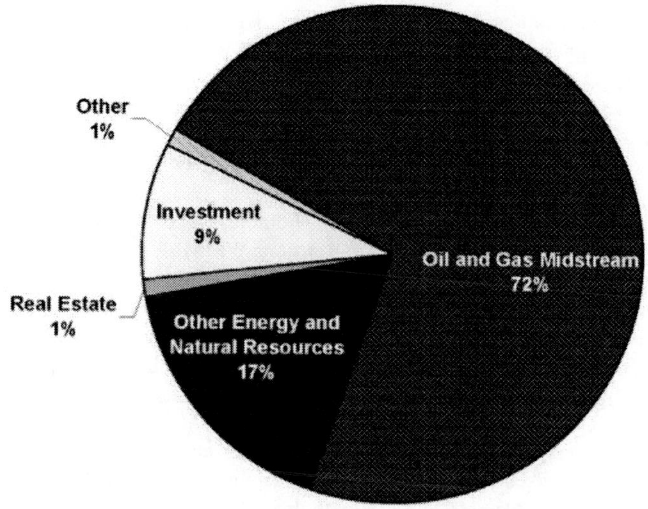

Source: CRS graphic using data from the National Association of Publicly Traded Partnerships, available at http://www.naptp.org/documentlinks/ Investor_ Relations/MLP_101.pdf.

Notes: Real estate includes income properties, homebuilders, and mortgage securities.

Figure A-2. MLPs by Market Capitalization, 2010.

End Notes

[1] The White House, "Remarks by the President in State of Union Address," press release, January 25, 2011, http://www.whitehouse.gov/the-press-office/2011/01/25/remarks-president-state-union-address.

[2] For additional background on other tax-favored financing options for renewable energy, see CRS Report R41573, Tax-Favored Financing for Renewable Energy Resources and Energy Efficiency, by Molly F. Sherlock and Steven Maguire.

[3] Technically, IRC § 7704(b) defines a partnership as an MLP if (1) "interests in such partnership are traded on an established securities market," or (2) "interests in such partnership are readily tradable on a secondary market (or the substantial equivalent thereof)."

[4] Other "pass-through" forms of business include sole proprietorships, partnerships, S corporations, and limited liability companies (LLCs). For more information on the various business forms, see CRS Report R40748, Business Organizational Choices: Taxation and Responses to Legislative Changes, by Mark P. Keightley.

[5] C corporations get their name from Subtitle A, Chapter 1, Subchapter C of the Internal Revenue Code, which partly determines the taxation of corporations.

[6] There are a variety of other forms of partnerships, including general partnerships, limited liability partnerships, and electing larger partnerships. For more information on the

differences in partnerships, see CRS Report R40748, Business Organizational Choices: Taxation and Responses to Legislative Changes, by Mark P. Keightley.

[7] See IRC § 7704(c). Another exception exists under IRC § 7704(g) for partnerships that were publicly traded on December 17, 1987. These partnerships may elect to retain partnership status by paying a tax equal to 3.5% of gross income from the active conduct of business as long as the line of business is unaltered.

[8] Interest generated from financial and insurance related businesses is not considered qualifying income for the purpose of MLP tax treatment.

[9] Wells Fargo Securities , MLP Primer - Fourth Edition, November 19, 2010, p. 15, http://www.naptp.org/ documentlinks/Investor_Relations/WF_MLP_Primer_IV.pdf.

[10] See the "Legislative History" section for more information.

[11] A separate operating company mitigates this problem because the operating company's owner (the MLP) never changes.

[12] Passive income is defined as income from trade or business activities in which the taxpayer does not materially participate.

[13] IRC § 469(k).

[14] Some of the gain that a taxpayer may realize from the sale of an asset could be the result of depreciation deductions taken in previous years. The gain that is attributable to depreciation is generally "recaptured" by taxing it according to the ordinary income tax rates rather than the more favorable capital gains rates.

[15] Internal Revenue Service, Tax on Unrelated Business Income of Exempt Organizations, March 2010, p. 8, http://www.irs.gov/pub/irs-pdf/p598.pdf.

[16] For additional background, see CRS Report RS22689, Taxation of Hedge Fund and Private Equity Managers, by Mark Jickling and Donald J. Marples.

[17] Apache Petroleum, formed in 1981, has been cited as the first MLP. See Jane R. Livingstone and Thomas C. Omer, "Publicly Traded Partnerships, Tax Cost, and Choice of Entity," Tax Notes, July 27, 2009, pp. 365-378.

[18] The Revenue Act of 1987 (P.L. 100-203).

[19] Technically, the tax treatment of MLPs is determined by the tax laws for publically traded partnerships (PTPs). PTPs are a more general form of partnerships, but the two terms are often used interchangeably. The National Association of Publicly Traded Partnerships, the trade association representing MLPs, provides the following discussion on the differences between PTPs and MLPs: "For most purposes, they [PTPs] are the same as MLPs, as the two terms are used interchangeably to refer to a publicly traded business entity which has chosen to be treated as a partnership under the tax laws. Technically, however, they are not quite the same thing. The term 'master limited partnership,' or MLP refers to a tiered limited partnership structure (i.e., a general partner manages the partnership and limited partners contribute capital) in which operations are conducted by lower-tier partnerships or other subsidiaries held by the publicly traded, 'master' limited partnership. Not all MLPs are PTPs—while most are publicly traded, a few are not. And, not all PTPs are really MLPs, even though they may be referred to as such. Some PTPs are actually publicly traded limited liability companies (LLCs) that have chosen to be taxed as partnerships. LLCs do not have a general partner, and investors have greater rights vis à vis management than in a limited partnership." For more information, see http://www.naptp.org/ PTP101/FAQs.htm.

[20] H.Rept. 100-391, 100th Cong., 1st Sess. 1065.

[21] H.Rept. 100-391, 100th Cong., 1st Sess. 1066.

[22] The Taxpayer Relief Act of 1997 (P.L. 105-34).

[23] Specifically, the law expands the definition of qualifying income to include income arising from the transportation or marketing of industrial source carbon dioxide.

[24] See U.S. Congress, Joint Committee on Taxation, General Explanation of Tax Legislation Enacted in the 110th Congress, committee print, 110th Cong., March 2009, JCS-1-09, p. 599.

[25] Sen. Tom Harkin, Statements on Introduced Bill and Joint Resolutions, Congressional Record, daily edition, July 21, 2008, pp. S6973-S6974.

[26] President Obama, in the 2011 State of the Union address, stated, "We'll invest in biomedical research, information technology, and especially clean energy technology—an investment that will strengthen our security, protect our planet, and create countless new jobs for our people." The text of this address is available at http://www.whitehouse.gov/the-press-office/2011/01/25/remarks-president-state-union-address.

[27] Domestic energy security is closely linked to petroleum import dependence. The majority of imported petroleum used in the U.S. is used in the transportation sector. Thus, from this perspective, enhancing domestic energy security will require technologies that reduce reliance on petroleum for motor vehicles. Thus, if the MLP structure can be used to encourage investment in renewable fuels, that ultimately reduce reliance on imported petroleum, the policy could arguably be one that is consistent with domestic energy security objectives.

[28] For more information on "green jobs," see CRS Report R40833, Renewable Energy—A Pathway to Green Jobs?, by Richard J. Campbell and Linda Levine. The potential for green job creation has been questioned by others. One reason for skepticism stems from the possibility that many of the technologies being installed involve components that are primarily manufactured overseas. For additional background on some of the uncertainty surrounding the green job potential, see Michael J. Graetz, The End of Energy: The Unmaking of America's Environment, Security, and Independence (Cambridge, MA: The MIT Press, 2011), pp. 168-171.

[29] Renewable energy projects can monetize (monetization is simply the conversion of future tax benefits into cash), at a discount, federal tax incentives (PTC, ITC, and accelerated depreciation) through an investor for cash that can be used to pay for a portion of project capital and installation costs.

[30] For additional background, see CRS Report R41635, ARRA Section 1603 Grants in Lieu of Tax Credits for Renewable Energy: Overview, Analysis, and Policy Options, by Phillip Brown and Molly F. Sherlock.

[31] See CRS Report R41635, ARRA Section 1603 Grants in Lieu of Tax Credits for Renewable Energy: Overview, Analysis, and Policy Options, by Phillip Brown and Molly F. Sherlock, Mark Bolinger, Ryan Wiser, and Naim Darghouth, "Preliminary evaluation of the Section 1603 treasury grant program for renewable power projects in the United States," Energy Policy, vol. 38, no. 11 (November 2010), pp. 6804-6819, and Paul Dickerson, "The (Too Short) Extension of Section 1603 Renewable Energy Cash Grants," The Electricity Journal, vol. 24, no. 2 (March 2011), pp. 27-33.

[32] Nate Gorence and Sasha Mackler, Reassessing Renewable Energy Subsidies: Issue Brief, Bipartisan Policy Center, Washington, DC, March 22, 2011, http://www. bipartisanpolicy. org/sites/default/files/BPC_RE%20Issue%20Brief_3- 22.pdf.

[33] Ibid.

[34] The potential value of MLPs to the renewable energy sector was recognized before the financial crisis and the enactment of the Section 1603 grant program (see Keith Martin, "Master Limited Partnerships," Project Finance Newswire, March 2006). With the

expiration of the Section 1603 grant program at the end of 2011 approaching, there has been increased interest in looking at alternative policies for enhancing investment and facilitating financing of clean energy technologies.

[35] Proposals that would expand the definition of qualifying income to include revenues from renewable energy activities might also include provisions that would change current passive activity loss limitations to allow losses generated from renewable energy tax incentives to offset other income.

[36] Nate Gorence and Sasha Mackler, Reassessing Renewable Energy Subsidies: Issue Brief, Bipartisan Policy Center, Washington, DC, March 22, 2011, http://www.bipartisan policy.org/sites/default/files/BPC_RE%20Issue%20Brief_3- 22.pdf.

[37] See U.S. Congress, Joint Committee on Taxation, Present Law And Analysis Relating To Tax Treatment Of Partnership Carried Interests And Related Issues, Part I, committee print, 110th Cong., September 4, 2007, JCX-62-07, p. 40.

[38] As noted above, the modification to the definition of qualifying income under EESA resulted in an estimated revenue loss of $119 million over 10 years. Ultimately, the size of the revenue loss will depend on what types of renewable and clean energy activities are deemed qualifying income. Further, if policies extending the MLP structure to renewables are coupled with changes in the passive activity loss rules, additional federal revenue losses may result.

[39] U.S. Congress, Joint Committee on Taxation, Estimates of Federal Tax Expenditures for Fiscal Years 2010 - 2014, committee print, 111th Cong., December 15, 2010, JCS-3-10, p. 37.

[40] Nicola M. White and Drew Pierson, "Baucus Says Congress Should Look at Taxing Passthroughs as Corporations," Tax Notes Today, May 5, 2011.

[41] Statement made during a Senate hearing by Treasury Secretary Timothy Geithner. U.S. Congress, Senate Committee on Finance, The President's Budget for Fiscal Year 2012, 112th Cong., February 16, 2011.

[42] For additional background, see Michael J. Graetz, The End of Energy: The Unmaking of America's Environment, Security, and Independence (Cambridge, MA: The MIT Press, 2011), pp. 124-126.

[43] When tax incentives for wind were reintroduced in 1992 the incentive was designed to reward investment, rather than production.

[44] See Bloomberg New Energy Finance, Crossing the Valley of Death: Solutions to the Next Generation Clean Energy Project Financing Gap, June 21, 2010.

[45] See Wells Fargo Securities, MLP Primer, 4th Ed., November 19, 2010, http://www.naptp. org/documentlinks/ Investor_Relations/WF_MLP_Primer_IV.pdf.

[46] Ibid.

INDEX

A

B

C

D

E

F

G

H

I

J

L

Q

R

S

T

U

V

W

Y